Electrophysiology of the Neuron

ELECTROPHYSIOLOGY OF THE NEURON

An Interactive Tutorial

JOHN HUGUENARD
DAVID A. McCORMICK

A Companion to *Neurobiology*
by Gordon Shepherd

New York   Oxford
OXFORD UNIVERSITY PRESS
1994

Oxford University Press

Oxford   New York   Toronto
Delhi   Bombay   Calcutta   Madras   Karachi
Kuala Lumpur   Singapore   Hong Kong   Tokyo
Nairobi   Dar es Salaam   Cape Town
Melbourne   Auckland   Madrid

and associated companies in
Berlin   Ibadan

Library of Congress Cataloging-in-Publication Data
Huguenard, John.
Electrophysiology of the neuron : an interactive tutorial / John
Huguenard and David A. McCormick.
p. cm.   Companion to: Neurobiology by Gordon Shepherd.
This manual contains a program for simulating the generation of
different patterns of activity in single neurons on an IBM-AT
compatible computer. What is needed: An IBM-AT compatible computer
(8086, 80286, 80386, or 80486 processors) with or without a math co
-processor and with CGA, EGA, or VGA graphics in either color or
monochrome. A hard disk drive is highly recommended, but not
required.
Includes bibliographical references.
ISBN 0-19-509111-6
1. Neurophysiology — Computer simulation — Laboratory manuals.
2. Neurons — Computer simulation — Laboratory manuals.
3. Electrophysiology — Computer simulation — Laboratory manuals.
I. McCormick, David.   II. Shepherd, Gordon M., 1933 — Neurobiology.
III. Title.
QP357.H84   1994
612.8 — dc20   93-39004

579864
Printed in the United States of America
on acid-free paper

Contents

PC Version

This manual contains a program on a 3½″ double-sided, double-density (720 K) floppy disk for simulating the generation of different patterns of activity in single neurons on an IBM-PC–compatible computer. Necessary equipment includes an IBM-PC–compatible computer (8086, 80286, 80386, or 80486 microprocessors) with or without a math coprocessor and with CGA, EGA, or VGA graphics in either color or monochrome. A hard disk drive is highly recommended but not required.

Introduction

When neurophysiologists study the electrophysiology of neurons and neuronal interactions, the recordings occur in real time, which imparts to the experimenter a level of excitement and insight that is difficult to communicate through static figures alone. This manual and associated computer program will allow you to share in that experience. First, you will repeat the classical experiments of Hodgkin and Huxley, and learn the ionic basis of action potential generation in the squid giant axon. Later, the inclusion of more recently discovered intrinsic and synaptic currents will allow you to investigate the electrophysiological properties of more complex excitable cells, including sympathetic ganglion cells, hippocampal and cortical pyramidal cells, and thalamic relay neurons. Perhaps in doing your own simulations you will share a bit of the excitement that the original researchers felt in the discovery of the different ionic components that enable different cell types to behave in their own unique manner in accordance with their varying roles in neural function.

The following tutorial is designed to allow you to "discover" for yourself the mechanisms of generation of electrical activity in different preparations through the performance of "experiments" on your computer. In this way, you will be able to control certain experimental conditions, such as the intensity and direction of intracellular current injection or the alteration of intracellular and/or extracellular ionic concentrations, in order to examine the effects of these experimental manipulations on the responses of the cell. In another set of investigations, voltage-clamp experiments will be performed to study the time- and voltage-dependent properties of ionic currents that lead to the patterned responses observed in neurons. Each example is carefully integrated with examples from the original experiments, most of which are reviewed in Chapters 4, 5, and 7 of Gordon Shepherd's textbook, *Neurobiology* (1994). Thus, this manual gathers in one place a

3

dynamic description of neuronal properties, so that you can better understand those properties as you encounter them in the different systems covered throughout Shepherd's *Neurobiology*.

This manual is organized into several parts. The beginning student of electrophysiology should be sure to perform the experiments in the sections "Resting Potential," "Action Potential," and "Synaptic Potentials." The experiments in the section "Neurophysiological Properties of Neurons" extend these basic electrophysiological experiments by adding additional ionic currents that have been discovered in neurons. For those who are interested, the mathematical methods of Hodgkin and Huxley in the modeling of ionic currents, Appendix B, "A Brief Explanation of How the Model Works," is also included.

If you don't have a computer at hand, don't despair, for we will present the findings in figures as well, thereby allowing you to follow along without the aid of a computer.

Installation on an IBM-AT–Compatible Computer

Necessary Equipment

An IBM PC or AT or compatible computer with either Hercules, CGA, EGA, VGA, or 8514A graphics display in either monochrome or color. The faster the computer, the faster the simulations will run. In addition, the presence of a math co-processor also increases the speed of simulation. We recommend a computer with a 80386 or higher central processing unit (CPU) and a math co-processor, although these are not necessary and we have attempted to make the simulations quick enough to enjoy even on the slowest of computers! The simulation program runs relatively slowly from floppy drives; therefore, we strongly recommend installing it onto your hard disk. It will require approximately 500K of disk space[1] and a minimum of 256K of available random access memory (RAM). We have successfully run this modeling program on computers based on the Intel 8086, 80286, 80386 and 80486 microprocessors with either Hercules, CGA, EGA, or VGA graphics displays.[2]

At this point, let us mention a couple of conventions for this manual. **Bold type** indicates characters to be typed by you on your computer keyboard in order to perform the experiments. *Words in italics* are display items that appear on your computer screen. Finally, all commands, except pressing command keys (such as **O** for open file and **B** for begin simulation, and so on) are to be followed by depressing the Return or Enter key, as usual.

---

[1] For systems with only 360K disk drives: It is possible to store the programs on two separate floppy disks, one with the voltage-clamp program, and one with the current-clamp program.
[2] If you experience technical problems, please write to John Huguenard, Department of Neurology, Stanford University School of Medicine, Stanford, CA 94305; or David A. McCormick, Section of Neurobiology, Yale University School of Medicine, 333 Cedar St., New Haven, CT 06510. Give exact details on the computer you are using and the problem that is occurring.

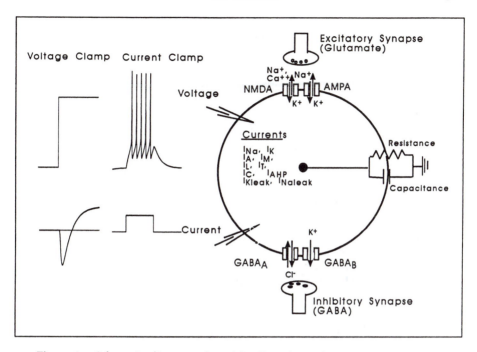

Figure 1. Schematic diagram of model cell. Voltage-clamp experiments can be performed by typing V-CLAMP. Current-clamp experiments can be performed by typing C-CLAMP. The modeled neuron has a number of different ionic currents, an excitatory synapse that uses AMPA and NMDA receptors and an inhibitory synapse that uses $GABA_A$ and $GABA_B$ receptors, and passive resistive and capacitive properties.

To install the programs, please do the following:

1. Insert the floppy disk containing the program and change the default drive to the one containing the floppy disk. For example, if you insert the floppy disk containing the modeling program into the A: drive and your hard disk is the C: drive (which is typically the case), then at the C:\> prompt type the following:
   A: (followed by a Return).

2. Now at the A:> prompt, type **INSTALL C: \NEURON COLOR** (be sure to leave a space between C: and \NEURON and between \NEU-RON and COLOR; if you are using a monochrome screen, type **IN-STALL C: \NEURON MONO**).[3] This command will call an installation routine on the floppy disk that will creat a subdirectory called NEURON (or whatever you type after the \) and will unpack all of the files needed to run the program and place them in this subdirectory.

---

[3]If you answer color for a monochrome screen, some of the traces will blink on and off continuously during the simulations. To change to monochrome screen after installation is complete type **setmono.bat**. To change to a color screen type **setcolor.bat**.

Once the installation routine is complete, control will be shifted to the subdirectory in which you installed the program. You should now see C:\NEURON> on your computer screen. You are now ready to do experiments!

Using an IBM-Compatible Computer to Perform the Two Basic Types of Experiment

There are two basic types of electrophysiological experiment (see Chapter 4 in *Neurobiology*). The first is to inject current into a cell and record what subsequently happens to the voltage. This type of experiment is typically referred to as a *current-clamp* experiment, since the amount of current that you inject is held constant (i.e., is "clamped") by the intracellular recording amplifier (Figure 1). A fundamentally different type of experiment is performed when you vary the amount of current you inject into the cell in order to hold the membrane potential (voltage) constant. This type of experiment is referred to as a *voltage-clamp* experiment (Figure 1). Similarly, there are two modeling programs. For performing current-clamp experiments, you run the program by typing **C-CLAMP** from the DOS prompt (e.g. C:\NEURON>). For performing voltage-clamp experiments, you type

---

CCLAMP Ver. 3.2 Begin
Text comment Open (read) **Keep** (write) **Quit Replot Y:Overlay**
Comments: resting potential

Trial duration (ms) = 150	Starting Vm (mV) = -65
Base current (nA) = 0	
Onset of step (ms) = 50	onset of ipsc = 200
Offset of step = 100	gIPSP (0-0.1) = 0
Injected current (nA) = 0	onset of epsc = 500
	gEPSP (0-0.1) = 0
gNa (0-20 µS) = 0	
gK (0-4) = 0	mM [Mg]o (1) = 1
pT (0-40 pL/sec) = 0	mM [Ca]o (2) = 2
pL (0-100) = 0	nM [Ca]i (10) = 10
gC (0-2) = 0	mM [K]o (3.1) = 3.1
gA (0-2) = 0	mM [K]i (135) = 135
gM (0-0.1) = 0	mM [Na]o (145) = 145
gAHP (0-0.1) = 0	mM [Na]i (31) = 31
pKleak (0-1) = 1	mM [Cl]o (120) = 120
pNaleak (0-0.1) = 0.06	mM [Cl]i (7) = 7

Figure 2. C-CLAMP menu screen. This screen allows you to enter the parameters of the current-clamp experiment including the injection of a current pulse, the *maximal* conductances of the different ionic conductances (e.g., gNa), the membrane potential at which the simulation begins, the onset and amplitude of an inhibitory and excitatory postsynaptic potential, and the extracellular and intracellular concentrations of the different ions.

V-CLAMP from the DOS prompt. After starting either C-CLAMP or V-CLAMP you will see a menu on your screen (Figures 2 and 8). By moving the cursor through these menus with the arrow keys on your keyboard you can determine some of the various parameters of the experiment, including values of the various ionic conductances and the concentrations of ions inside and outside the model cell. To assist you in getting started we have included reasonable suggested values for each of these parameters on the menu screen.

This manual covers the resting potential, the action potential, neurophysiological properties of neurons, and synaptic responses, which include material from Chapters 4, 5, and 7 in *Neurobiology*. In each section we begin with experiments performed in the current-clamp mode of recording and follow with voltage-clamp experiments, as did the original investigators. We will also present figures from some of the original experiments, so that you may compare your results with theirs.[4] If installation was successful, you may now proceed to each of these sections.

---

[4]The results of your simulations will resemble, but not be identical to, those of the original investigators, since these were obtained from diverse biological systems under various conditions that are not included here.

Resting Potential

The presence of a potential difference (voltage) between the inside and outside of axons and neurons at rest (the resting potential) is essential for the generation of electrical activity. Consequently, understanding the mechanisms of generation of this resting membrane potential is also fundamental to understanding the mechanisms of electrophysiological excitability in neurons and axons. Therefore, in this section we will perform experiments on the cellular mechanisms of determination of the resting potential.

Alan Hodgkin, Andrew Huxley, and Bernard Katz, working at the Plymouth Biological Laboratory in England, took advantage of the large size (approximately 1 mm in diameter) of the squid giant axon[5] to perform intracellular recordings of action-potential generation and to examine the ionic dependence of this (Hodgkin & Katz, 1949). To perform experiments similar to theirs, start the C-CLAMP program by typing **C-CLAMP** at the DOS prompt C:\NEURON> follwed by a Return. Your computer screen should now appear as in Figure 2.

Now press the **O** key to open a parameter file. Using the up arrow and down arrow keys on the right hand side of your keyboard, move the highlighted file name until REST.CCS appears. Now press the Return or Enter key to select this parameter file. You can now begin the simulation by pressing the **B** key.

This simulation demonstrates the resting membrane potential of the squid giant axon (Figure 3; Na and K leak). Note that the resting potential is -65 mV, which is in-between the reversal (equilibrium) potentials for both K^+ (-100 mV) and Na^+ ($+41$ mV). The resting membrane potential is determined by the leakage, at rest, of ions across the membrane. If ions are in differing concentrations across the membrane and are able to pass

---

[5]Large axons conduct quickly and therefore are often used for generating escape reflexes in invertebrates, as is the case for the giant axon in the squid.

through the membrane, they diffuse down their concentration gradient. However, in so doing they carry charge with them, thereby changing the membrane potential. This change in membrane potential will also have an effect on the distribution of the ions, since, as you remember from high school science, like charges (e.g., positive and positive) repel, and unlike charges (e.g., positive and negative) attract. When the influences of the concentration gradient and membrane potential perfectly balance each other, the ion is said to be "in equilibrium." The value of membrane potential that is able to perfectly balance the difference in concentration gradient (e.g., −100 mV is needed to keep the higher concentration of K^+ inside the cell from leaking out) is referred to as the "equilibrium potential" and can be calculated using an equation developed by W. Nernst and appropriately known as the Nernst equation. A modification of the Nernst equation was developed by Goldman, Hodgkin, and Katz and is known

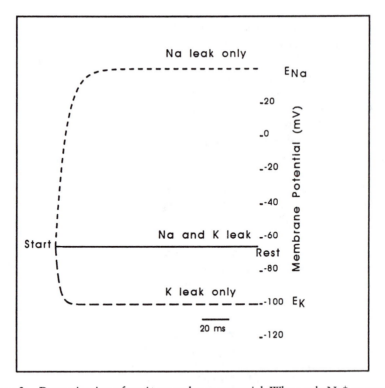

Figure 3.   Determination of resting membrane potential. When only Na^+ can move across the membrane, the membrane potential moves to E_{Na} (+41 mV). When only K^+ can transverse the membrane, the membrane potential moves to E_K (−100 mV). When the membrane is conductive to both Na^+ and K^+, the membrane potential comes to a point in-between E_K and E_{Na}, but closer to E_K.

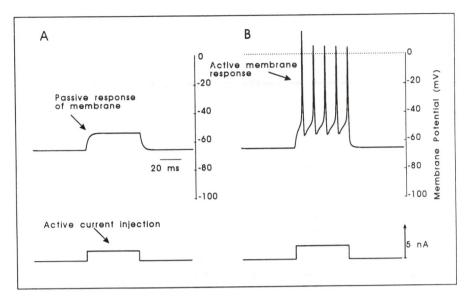

Figure 4. Passive and active membrane responses of model axon. *A,* Injection of a 1.5-nA current pulse (indicated by *lower trace*) results in only a passive response that results from the resistive/capacitive properties of the membrane. *B,* Injection of 2.0 nA of current results in a series of action potentials.

as, you guessed it, the Goldman-Hodgkin-Katz(GHK) constant field[6] equation (see Chapter 5 in *Neurobiology*). A brief explanation of these equations is given in Appendix A, "Nernst and Goldman-Hodgkin-Katz Constant Field Equations" at the end of this manual.

Experiment 1: Equilibrium Potential

In the model, the membrane is permeable or "leaky" to two different ions, Na^+ and K^+. These two permeabilities, *pKleak* and *pNaleak*, set the resting membrane potential of the cell, which is calculated using the GHK constant field equation (see Chapter 4 in *Neurobiology*). The term "leak" refers to the fact that the permeability of the membrane to these ions does not change with changes in membrane potential and contributes substantially to the resting "leakiness" of the membrane. To examine the influence of *pNaleak* and *pKleak* in the computer model, first press the **R** key (return) to get back to the menu screen (this is to be done after each simulation when you want to return to the menu screen) and use the arrow keys to move the cursor on your computer screen to *pNaleak=0.06* and replace 0.06 with 0. Now press the **B** (begin) key to do a simulation. Notice that

---

[6]One of the primary assumptions of the GHK equation is that the electrical field within the membrane is constant, and hence the name "constant field equation." See Chapter 4 in *Neurobiology* (Shepherd, 1994).

now the membrane potential starts at -65 mV, as we determined in the main menu (Figure 2), but then quickly falls to -100 mV, which is the equilibrium potential for K^+ ions (Figure 3; K leak only). Potassium ions are in much greater concentration on the inside of the cell than on the outside. When we start the simulation at -65 mV, K^+ ions flow down this concentration gradient (i.e., out of the cell). However, as they exit the cell, the membrane potential becomes more negative and this generates a "pull" on the K^+ ions to not leave the cell. As mentioned above, when the tendency for K^+ to exit the cell down its concentration gradient perfectly balances the tendency for K^+ ions to enter the cell due to the negative charge inside the cell, the ion comes into equilibrium and the membrane potential is equal to the equilibrium potential. Note that although there is a net movement of ions across the membrane when the membrane potential is away from the equilibrium potential for this ion, the number of ions that make this trip is actually very small in comparison to the number of ions in the cell and in the bathing medium and therefore the concentrations on both sides of the membrane typically change by only a negligible amount.[7]

Experiment 2: Effects of Changing Ion Concentrations

You can examine the effects of changing the extracellular or intracellular ion concentrations on the equilibrium potential yourself. Press the R key to get back to the menu screen and experiment with the extracellular K^+ concentration, $[K^+]_o$. For example, use the arrow keys to move to $[K^+]_o$ and set it to be 135 so that it is equal to the intracellular concentration ($[K^+]_i$). Now press the Y (overlay) key to obtain an overlay of the present simulation with the last. Note that now the membrane potential settles at 0 mV, since there is no transmembrane concentration gradient for K^+ ions and therefore the ion will come into equilibrium when there is no voltage difference across the membrane. To examine the membrane potential with only a leak Na^+ conductance, press the O (open) key and reload the REST.CCS file. Press the B (begin) key to repeat the experiment under "control" conditions (i.e., with normal resting Na^+ and K^+ leak). Now go to pKleak=1 and change this to 0 to turn off the contribution of K^+ to the membrane potential. Press Y (overlay) and note that the membrane potential now comes to rest at the equilibrium potential for Na^+ (E_{Na}, $+41$ mV: Figure 3; Na leak only).

When more than one ion can move across the membrane, the membrane potential will come to rest somewhere between the equilibrium potentials of the different ions, depending upon the relative conductance or permeability of the membrane to each of the different ions. This can be examined

---

[7]When there are high levels of neuronal activity in real neural systems, the concentrations of some ions can change significantly in the region just inside or outside the membrane for a brief period of time. The free Ca^{++} concentration inside the cell is kept very low (approximately 50–100 nM) through special buffering mechanisms. It may rise into the μM range during activity, but is quickly buffered thereafter.

here by changing *pNaleak* and *pKleak*. Move the cursor to *pKleak=0* and change this to **10**. Press the B key and note that the membrane potential now comes much closer to E_K (-100 mV) than it did when *pKleak* was only 1. As pointed out in Chapter 4 of *Neurobiology*, normally the resting potential is determined mainly by the distribution of K^+ because the membrane is more permeable to K^+ than to Na^+. We have approximated this situation in the program by setting resting Na^+ permeability to only a fraction of resting K^+ permeability.

Experiment 3: Passive Membrane Properties

With the use of specialized electronic amplifiers, one can not only record the membrane potential in axons and neurons, but also can directly manipulate the membrane potential by injecting current. The resulting responses arise from both the passive properties of the plasma membrane and the active properties of the ionic channels located in this membrane. Using the O and B keys, load and run the file called PASSIVE.CCS. With the injection of a small "square' pulse of current, only the passive components are seen, and these appear as a change in membrane potential that exhibits "rounded corners" (Figure 4A). The slowed rise and fall of the membrane potential with the sudden onset (or offset) of current injection is due to the fact that the membrane acts as a resistance and capacitance in parallel (resistance is the inverse of conductance; Figure 1). Capacitance is the ability to store charged ions. When you inject current into the cell, at first most of this is used to "charge the capacitance" (which basically amounts to changing the distribution of ions near the cellular membrane). As the capacitance becomes charged, and you continue to inject current, the ion concentration gradients change a little and therefore the membrane potential changes.[8] How much does the membrane potential change? There is a known relation between voltage, current, and resistance known as Ohm's law, which is $V = I \cdot R$, where V is voltage in volts, I is current in amperes, and R is resistance in ohms. The inverse of resistance (measured in ohms) is conductance (which is measured in mhos or siemens and abbreviated as g) and mathematically is $g = 1/R$. In this way, Ohm's law with conductance is $V = I/g$. We can use these relationships in model or real cells to calculate the "input" resistance or conductance of a cell by measuring the voltage deviation generated at steady state by injection of a known amount of steady current and using Ohm's law.

Let's consider these terms just a bit more. Current is a measure of the rate of flow of charge through a conductor, such that 1 ampere is equal to 1 coulomb (C) of charge per second. Since a coulomb of charge is 6.25×10^{18} electrons, this amounts to 1 ampere being the passage of 6.25×10^{18} electrons per second through a conductor. Voltage, named after Alessandro

---

[8]For a more detailed discussion of membrane resistance and capacitance, see Kuffler et al. (1984).

Volta, is the "pressure" put on that charge such that 1 volt will move 1 ampere of charge through a 1-siemen conductor (which is equivalent to a 1-ohm resistor), as shown by Ohm's law. Conductance in biological membranes is directly related to the permeability of the membrane to the ion of interest.

It may be useful to consider a simple analogy. If we compared electricity to a simple water delivery system with an elevated storage tank and a delivery hose, current would be analogous to the amount of water running through the hose every second, and voltage would be analogous to the pressure placed on the water, which is directly related to the height of the tank. Conductance (and permeability) then would be related to the size of the hose. In this way, increasing the pressure (i.e., raising the height of the tank) or increasing the conductance (i.e., increasing the size of the hose) would result in an increase in the current (volume of water per second)![9]

Study Questions: Determination of Resting Membrane Properties

1. What are the two main factors that influence the movement of ions across the membrane?
2. What is meant by the "equilibrium potential" of a particular ion?
3. Why is the K^+ equilibrium potential -100 and the Na^+ equilibrium potential $+41$ when both are positive ions?
4. If the resting potential is -65 under normal conditions, what would it be if you reversed the concentrations of K^+ and Na^+ across the membrane? Load and run REST.CCS and reverse the concentrations of K^+ and then reverse the concentrations of Na^+ to see if you are correct.
5. Load and run PASSIVE.CCS. What would you expect to happen to the membrane potential and response to a depolarizing current pulse if you doubled both *pKleak* and *pNaleak*? Try this now and see if you are correct. Briefly explain the result.

Answers are on page 69.

---

[9] An amplifier that injects a constant current into a cell, as in current clamp, adjusts the voltage applied to the electrode in order to keep the current constant.

Action Potential

Neurons and their axons do much more than merely exhibit a resting potential. They also actively generate and propagate action potentials through their membrane, allowing communication from one part of the cell to another and, subsequent to transmitter release, communicate to other cells. Like the resting potential, the mechanisms of generation of the action potential are essential to understanding of the electrophysiological properties of excitable cells.

Experiment 4: Active Membrane Properties

With the injection of a larger current pulse, you may surpass the threshold for generation of action potentials. To do this while the parameter file PASSIVE.CCS is loaded, move the cursor with the up arrow and down arrow keys to the value of *Injected current (nA) = 1.5* and change *1.5* to *2.0*. Now press the **B** (begin) key. The larger injected current pulse depolarizes the axon beyond the threshold for action potentials (Figure 4B).

Notice that the peaks of the action potentials "overshoot" 0 mV, indicating that they are generated not by a simple nonselective increase in membrane permeability to all ions, which would have a reversal potential around 0 mV, but rather something more complicated, such as selective permeability changes.

Experiment 5: Effects of Different Ions on Impulse Generation

To investigate which ions are important in the generation of the action potential, you can change the concentration of different ions in the artificial "seawater" (which bathes the extracellular surface of the axon) one at a

15

time and observe the effects.[10] Among various ions in seawater, you know that there are significant amounts of Na^+, K^+, Ca^{++}, Mg^{++}, and Cl^- (see Chapter 4 in *Neurobiology*.[11] Move the cursor to the extracellular concentration of Mg^{++} *([Mg]$_o$=)*, and change this to 0.1[12] by typing 0.1 and again inject the current pulse into the cell and compare this to your last experiment by pressing the Y (overlay) key. With this experiment you notice that reducing Mg^{++} in the bathing medium has no effect on either the resting membrane potential or the generation of action potentials. Repeat this experiment with the other ions by changing $[Na^+]_o$, $[K^+]_o$, $[Cl^-]_o$ and $[Ca^{++}]_o$ to a low value (0.1 mM) one at a time and press the B key to rerun the simulation after each change. Remember, you can always restore the original values by loading the file ACTIVE.CCS with the O (open) key after each simulation.

You will quickly recognize that the two ions that are of particular significance are Na^+ and K^+. When you reduce extracellular Na^+ to 0.1, you notice two effects. First, the membrane potential of the cell becomes significantly more negative (Figure 5A). This finding indicates that the passive influx Na^+ ions into the cell normally contributes to the resting membrane potential, causing it to be depolarized to -65 mV, and when this influx of Na^+ ions is removed or reversed, the membrane potential becomes substantially more hyperpolarized. To compensate for this hyperpolarization, you must inject current into the cell through your microelectrode to move the membrane potential back to -65 mV. To do this in the computer model, move the cursor to *Base current (nA) = 0* and change 0 to 2.25 to inject 2.25 nA of constant current into the modeled cell and press the B key to begin. Now you notice that action potentials are blocked, indicating that extracellular Na^+ is important for their generation (Figure 5B). Even if you double the amount of current injected to 4 nA, the cell still does not generate action potentials.[13]

Now, let's investigate K^+ ions. First, reload ACTIVE.CCS parameter file with the O key and run it with the B key. Now reduce $[K^+]_o$ to 0.1 mM by moving the cursor to *[K$^+$]$_o$=3.1* and typing in 0.1. and pressing Y. Again, you notice that this manipulation results in the resting membrane potential becoming substantially more hyperpolarized (as expected from the Nernst and GHK equations; see Appendix A), indicating that the leakage of K^+

---

[10]In real experiments, changes in ion concentrations would also change osmolality, so this would have to be compensated for through the addition of other ions that do not permeate the ionic channels under investigation.

[11]The concentrations used in the present simulation are based upon mammalian cerebrospinal fluid and not actual seawater. This is done to simplify the transition to mammalian cells later on.

[12]In real experiments it is not generally possible to remove all of a particular ion. In addition, the present simulation program does not use Goldman-Hodgkin-Katz constant field equation for many of the currents and therefore may yield nonphysiological effects at very low ion concentrations. See Chapter 4 of *Neurobiology* (Shepherd, 1994) and Appendix A, "Nernst and Goldman-Hodgkin-Katz Constant Field Equations."

[13]With large current injections you may notice "ripples" in the membrane potential. These result from performance compromises in the calculation procedure.

ions across the membrane helps to determine the resting membrane potential (Figure 5*C*), as we studied above in the section "Resting Potential." To compensate for the change in membrane potential, you inject steady current into the cell by changing *Base current (nA)* to **0.8** nA and again pressing the **B** key.

You now see that, unlike reducing $[Na^+]_o$, reduction of $[K^+]_o$ does not abolish action-potential generation, although you notice that the hyperpolarizing potentials that occur after each action potential are now larger than before (compare Fig. 5*A* and *D*)! This can be observed directly on

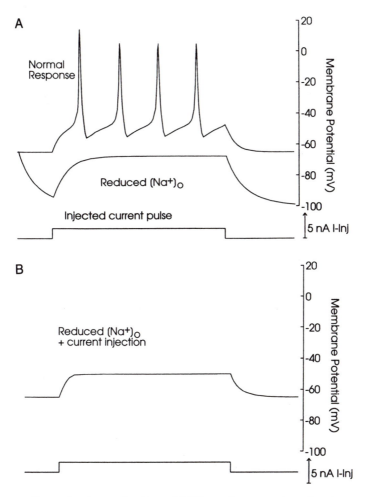

Figure 5. Effects of reducing $[Na^+]_o$ and $[K^+]_o$ on the generation of action potentials. *A*, Reduction of $[Na^+]_o$ to 0.1 mM results in hyperpolarization of the cell. *B*, Compensation for this hyperpolarization with the steady injection of current reveals that reduction of $[Na^+]_o$ has blocked the generation of action potentials, indicating that Na^+ is important for this active response. (*Figure continues on following page*)

your computer screen by now reloading ACTIVE.CCS with the O key and pressing the Y key. This increase in the hyperpolarization occurring after each action potential with reduced $[K^+]_o$ suggests to you that the movement of K^+ across the membrane must be important for this. In fact, you now consider two findings: reduction of $[K^+]_o$ results in an increase in the hyperpolarization after the action potential, and this hyperpolarization now goes below, or "undershoots," baseline membrane potential. Together, these events suggest to you that K^+ is in higher concentration inside the cell than out, and that K^+ ions move down this concentration gradient during the

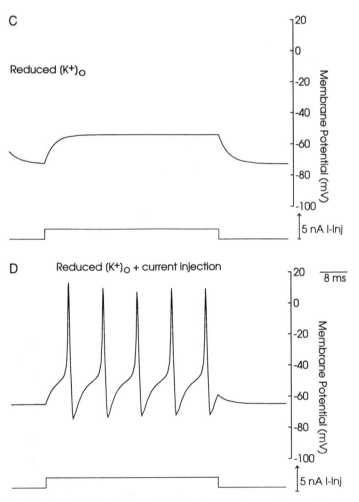

Figure 5. *Continued* C, Reduction of $[K^+]_o$ also results in a hyperpolarization of the cell; D, although now, compensation for this hyperpolarization with current injection reinstates action-potential generation. The enhancement of the hyperpolarizations after each action potential suggests that K^+ is important for the generation of these.

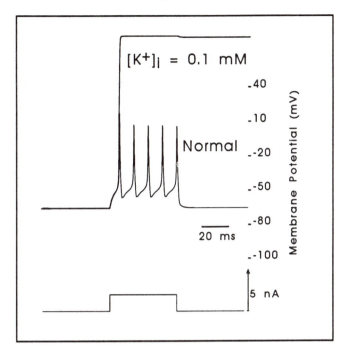

Figure 6. Reduction of intracellular K$^+$ concentration results in action potentials that do not repolarize.

repolarizing phase of an action potential! To test this possibility, you need also to change the intracellular concentration of K$^+$ ([K$^+$]$_i$).

Experiment 6: The Toothpaste Experiment

Baker, Hodgkin, and Shaw (1962) took advantage of the large size of the squid giant axon to squeeze out the axoplasm, as if the axon were a tube of toothpaste, and replace the axoplasm with an artificial one containing different ion concentrations. To replicate their experiments, press **O** load K_INTRA.CCS, and press **B**. You can reduce [K$^+$]$_i$ by changing this value in the model to **0.1 mM**. Now press **B**, and you find that the membrane potential depolarizes towards the equilibrium potential for Na$^+$ (which is +41 mV in these simulations). To compensate for this, you change *Base current (nA)* to −3 nA and press **B** again. Now when the axon generates an action potential, the action potential does not repolarize (Figure 6).[14] These results indicate that the outward movement of K$^+$ ions across the membrane potential is indeed critical to the repolarization of action potentials.

---

[14]The present computer model does not accurately replicate Na$^+$ channel inactivation in this regard. Normally, there may be some repolarization due to Na$^+$ channel inactivation.

Voltage Clamp

So now you have found that Na^+ and K^+—but not Cl^-, Mg^{++}, or Ca^{++} are critical to the generation of action potentials (in the squid axon) and you reason that Na^+ moves into the cell to depolarize it, while K^+ moves out of the cell to repolarize the membrane. However, you are not satisfied, since many questions remain; for example: Why is there generation of action potentials at all? Why isn't the flow of Na^+ into the cell just compensated for by the flow of K^+ out of the cell? Is the K^+ current perhaps delayed a bit? To understand this process in detail, Hodgkin and Huxley realized that they needed a scheme to describe in detail the movements of ions across the membrane (see Appendix B, "A Brief Explanation of How the Model Works," and also Chapter 5 in *Neurobiology*).

Examining the generation of action potentials closely, one realizes that the ionic currents involved have two main attributes in addition to their equilibrium potential: (1) a voltage dependence (i.e., the action potentials only occur above a particular membrane potential), and (2) a time dependence (the cell is first depolarized by Na^+ entry and then subsequently repolarized by K^+ departure). To get an accurate measure of these two factors, one would have to keep the voltage across the membrane constant and examine the amplitude and time course of the currents (ion fluxes) at different membrane potentials. In other words, one would have to perform voltage-clamp experiments, in which the membrane potential is fixed and time-dependent processes can be carefully measured in isolation.

Again, the large size of the squid giant axon comes in handy. Hodgkin and Huxley were successful in threading a metal wire through the axon and designing an electrical feedback circuit (at that time based upon vacuum tubes instead of transistors) that would adjust the amount of current injected into the axon in order to keep the voltage constant (see Figure 7). (Many students find voltage-clamp experiments difficult to follow. Let's

21

hope that the performance of interactive experiments with the present computer simulation can alleviate a bit of this confusion.)

Experiment 7: Voltage Clamp Analysis of Na⁺ and K⁺ Currents

To replicate Hodgkin and Huxley's voltage-clamp experiments, you must first quit the C-CLAMP program by pressing **Q** (quit) from the menu screen. To perform voltage-clamp experiments, type **V-CLAMP**. Your computer screen should now look like Figure 8.

Press the **O** key and load NA_K.VCS. Now press the **B** key and your screen should look like Figures 7A and 9. In this experiment you have moved the membrane potential from −100 mV to 0 mV and measured the amount of current you had to inject into the axon in order to keep this membrane potential constant which, in a closed system, is equal to the current flowing through the ionic channels. Notice that during this procedure, there is first an inward movement of ionic current (positive ions mov-

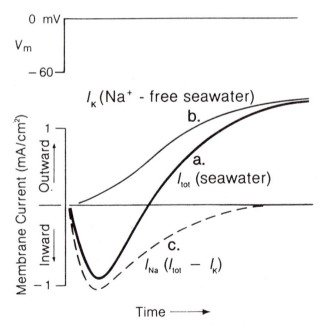

Figure 7. *A*, Response of the squid axon membrane under voltage clamp to a depolarization of 60 mV. *a*, Response in normal seawater. *b*, Response due to potassium current (I_K) when extracellular Na⁺ is replaced by choline ions. *c*, Calculated response due to Na⁺ current ($I_{Na} = I_{total} − I_K$). (From original study of Hodgkin & Huxley, 1952).

B PHARMACOLOGICAL BLOCKAGE

a. Control (I_{total})

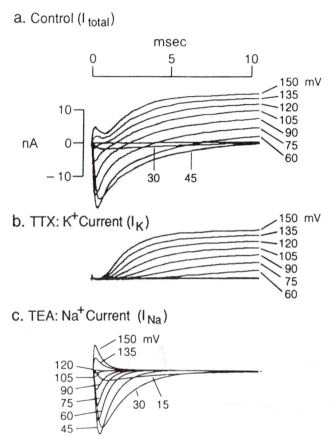

b. TTX: K⁺Current (I_K)

c. TEA: Na⁺Current (I_{Na})

Figure 7. *B*, Separation of ionic currents by use of nerve poisons. *a*, Response in normal seawater; different amplitudes of voltage steps are indicated on the right (in mV). *b*, Response due to I_K when I_{Na} is blocked by tetrodotoxin (TTX). *c*, Response due to I_{Na} when I_K is blocked by tetraethylammonium (TEA). (From Hille, 1977).

ing into the cell) followed by an outward movement of positive current (see Figure 9; solid line).

At this point, we need to define a bit of terminology that will be useful. In simple terms, ionic current through excitable membranes is controlled by two factors: (1) an ion-selective pore through which only certain ions can flow, and (2) a gate or gates that open(s) and close(s) the pore to allow ionic flux. The turning on of a current is known as the *activation* of the current and the opposite of activation is known as *deactivation*. These processes occur when an *activation gate* opens or closes. If a current turns on and then off despite a constant change in membrane potential, it is said to *inactivate*. The reverse of inactivation is *deinactivation*. Inactivation and

```
             VCLAMP Ver. 3.2  Begin  Text comment
        Open (read)  Keep (write)  Quit  Replot  Y:Overlay  Indiv. Steps


   Comments: K+ current only

   # of time steps = 200          gIPSP (0-0.1) = 0
   I max (nA) = 150               gEPSC (0-0.1) = 0
   I min (nA) = -200              IPSC onset (ms) = 120
   V max (mV) = 10                EPSC onset (ms) = 20
   V min (mV) = -100
                                  mM [Mg]o  (1) = 1
   gNa (0-20 µS) = 10             mM [Ca]o  (2) = 2
   gK (0-2) = 2                   nM [Ca]i  (10) = 10
   pT (pL/sec) (0-40) = 0         mM [K]o   (3.1) = 3.1
   pL (0-80) = 0                  mM [K]i   (135) = 135
   gC (0-2) = 0                   mM [Na]o  (145) = 145
   gA (0-2) = 0                   mM [Na]i  (30) = 30
   gM (0-0.1) = 0                 mM [Cl]o  (140) = 140
   gKleak (0-1) = 0               mM [Cl]i  (8) = 8
   gNaleak (0-1) = 0
```

Figure 8.   V-CLAMP menu screen. From here you can format the scale and offset
of the voltage and current displays, determine the maximal conductances for each
ion, and determine the extracellular and intracellular ion concentrations.

deinactivation occur when a separate *inactivation gate* closes or opens. For
current flow to occur, all activation and inactivation gates on a given pore
must be open. Some ionic currents, such as the Hodgkin and Huxley K^+
current, do not contain inactivate gates and therefore only activate (turn
on) and deactivate (turn off). Others, in contrast, contain both activation
and inactivation gates, and thus inactivate and deinactivate as well as activate and deactivate, such as the Hodgkin and Huxley Na^+ current.[15]

Experiment 8: Effects of Ion Concentrations on Ionic Currents

You can now perform the ion substitution experiments that you did during
current-clamp recordings by changing the values of $[Na^+]_o$, $[Na^+]_i$, $[K^+]_o$,
and $[K^+]_i$ (do this now; after each change reload Na_K.VCS with the O
key to restore the original values). During your experiments you find that
the inward current is sensitive to the concentration of Na^+ on either side
of the membrane, while the outward current is sensitive to K^+ concentrations (see Figure 9). For example, change the extracellular concentration of

---

[15]A simple analogy of activation and inactivation gates would be a tube in which you placed
one hand over each end. Removing one hand from the tube would correspond to activation,
while removing the other would correspond to deinactivation. You can see that the tube will
only allow objects to pass if both hands are removed (i.e., the pore is activated and not
inactivated).

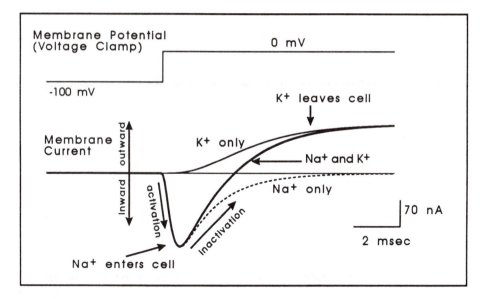

Figure 9. Voltage-clamp pulse delivered to a model squid giant axon. The membrane potential is held at −100 mV and then stepped to 0 mV. In response to this depolarization the Na⁺ current activates, allowing Na⁺ to flow into the cell, and then inactivates. The K⁺ current activates at a slower rate and allows K⁺ ions to leave the cell, thus causing an outward current. The individual ionic currents can be isolated by making the extracellular and intracellular concentrations of the other ion equal.

Na⁺ to match that of the intracellular concentration by typing **30** after $[Na^+]_o=$. Since in this experiment you are stepping to 0 mV (which is the equilibrium potential for ions that are in equal concentrations on both sides of the membrane), the Na⁺ current is eliminated and you see only the outward K⁺ current (Figure 9; thin line labeled K⁺ only). Now change $[Na^+]_o$ back to **145** and change $[K^+]_i=$ from 135 to **3.1** mM so that the K⁺ current is eliminated. Press B again and you will now see the Na⁺ current in isolation (Figure 9; dashed line; labeled Na⁺ only). Notice that the Na⁺ current and the K⁺ currents differ in a number of aspects. First, the Na⁺ current turns on (activates) more quickly. In other words, its *kinetics* are faster. Second, the Na⁺ current turns off (inactivates) even though the membrane potential is held constant, while the K⁺ current does not (Figure 9). This indicates that the Na⁺ current is *inactivating*, while the K⁺ current is *noninactivating* (i.e., persistent).

Experiment 9: Voltage Dependence of Na⁺ and K⁺ Currents

You may now wish to investigate the voltage dependence of the two ionic currents in more detail. To do this, you perform a number of voltage steps to different membrane potentials and examine the resulting currents. Press the O key and load NA_K_IV.VCS. Now press the I (individual steps) key once after each simulation. Each time you press this key, the membrane potential to which the axon is stepped is increased by 10 mV. In this manner, you can replicate the data obtained from the node of Ranvier in peripheral axons by Bertil Hille shown in Figure 7B. Let's now take advantage of pharmacological tools that selectively block the Na^+ and K^+ currents. Tetrodotoxin (TTX), isolated from the puffer fish, is a highly potent and selective toxin that blocks the Na^+ channels involved in action-potential generation (the puffer fish, *fugu*, is a Japanese delicacy that must be prepared with great care by the chef![16]). You can replicate the block of these channels by reducing the maximal conductance for Na^+ *(gNa (μS)=)* to 0. After blocking the Na^+ current, again repeatedly press the I key and note the amplitude and time course of the isolated, outward K^+ current (Figure 10). Now change *gNa(μS)=* back to 10 μS. Tetraethylammonium$^+$ (TEA) is an ion that blocks K^+, but not Na^+ channels. You can replicate this experiment by decreasing *gK=* to 0. Repeatedly press I again and note the voltage and time dependence of the now isolated Na^+ current (Figure 10). Congratulations—you have just replicated Hodgkin and Huxley's basic voltage-clamp experiments of the squid giant axon! (Don't, however, expect a call from Stockholm: the Nobel Prize for this discovery has already been awarded to Hodgkin and Huxley.)

By measuring the voltage dependence and kinetics of activation and inactivation of the Na^+ and K^+ components of the action potential, Hodgkin and Huxley were able to generate a kinetic scheme that allowed them to generate a mathematical model of the action potential (e.g., Figure 11; for a detailed explanation, see Appendix B, "A Brief Explanation of How the Model Works"). These authors proposed that the rapid upswing of the action potential was mediated by the entry of Na^+ ions into the axon in a positive-feedback manner such that the entry of Na^+ into the cell led to additional Na^+ entry through depolarization of the membrane. The fall of the action potential was proposed to be mediated by the *inactivation* of the Na^+ conductance and the slower or "delayed" activation of a K^+ current (which gave this K^+ current its name of "delayed rectifier") (Figure 11).

The computer model that you are now using makes use of mathematical equations similar to those of Hodgkin and Huxley, although they are based upon data obtained in mammalian thalamic relay neurons, cortical pyramidal cells, and sympathetic ganglion cells (Huguenard & McCormick, 1992; McCormick & Huguenard, 1992; Yamada et al., 1989). In the early 1950s Hodgkin and Huxley were forced to perform their calculations on a mechanical calculator in which the calculation of one action potential took an entire night of manual cranking! Luckily, students nowadays can generate similar calculations in a matter of seconds on their own computers.

---

[16]Tetrodotoxin may also be an active ingredient used by Caribbean witchdoctors to create zombies (through brain damage from anoxia) from their otherwise healthy victims.

Experiment 10: Analysis of Amplitude and Time Course of Individual Currents

Let us now take advantage of the computer to examine the amplitude and time course of the currents and the conductances involved in action-potential generation. To do this, first quit V-CLAMP by pressing **Q** from the menu screen and run the current clamp simulator (**C-CLAMP**) and load and run the file GNA_K.CCS with the **O** and **B** keys. The output will now show the spontaneous generation of a single action potential and the amplitude

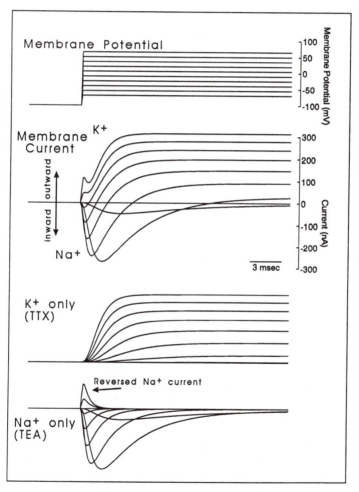

Figure 10. Full voltage range response of Na^+ and K^+ currents in the modeled squid axon. Stepping the membrane potential from -70 to $+80$ mV illustrates the activation of Na^+ and K^+ currents. Block of the Na^+ conductance (e.g., with tetrodotoxin [TTX]) reveals the outward K^+ current. Block of the K^+ conductance (e.g., with tetraethylammonium [TEA]) reveals the Na^+ current in isolation.

A

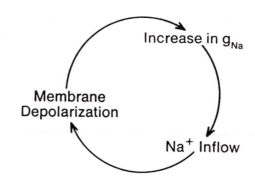

B

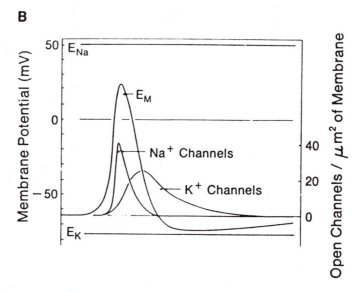

Figure 11. *A*, Regenerative relation between membrane depolarization, increase in Na$^+$ permeability and conductance, and Na$^+$ current that underlies the action potential. *B*, Reconstruction of changes in ionic conductance underlying the action potential according to Hodgkin-Huxley model; scale for the membrane potential (E$_M$) is shown on the left. The equilibrium potential for Na$^+$ and K$^+$ are also indicated on the left. Changes in Na$^+$ and K$^+$ conductances are scaled on the right in terms of calculated open channels per square micrometer of membrane. The time courses of these changes are controlled by a set of equations that constitutes the Hodgkin-Huxley model. (From Chapter 6, *Neurobiology*, Shepherd, 1994.)

and time course of the Na⁺ and K⁺ conductances (Figure 12). Note that the membrane potential depolarizes more and more as the Na⁺ conductance activates (Figure 12). The entry of Na⁺ into the cell depolarizes the membrane and this depolarization opens more Na⁺ channels. Thus, the Na⁺ conductance rapidly depolarizes the cell in a positive-feedback manner. The membrane potential then overshoots 0 mV as it attempts to reach the equilibrium potential of Na⁺ ions (+41 mV). However, at a somewhat slower pace, the K⁺ conductance is also increasing, allowing K⁺ ions to leave the cell, thus offsetting the inward flow of Na⁺ ions. This outflow of K⁺ ions prevents the action potential from reaching E_{NA}. *The inactivation of the Na⁺ conductance and the activation of the K⁺ conductance gives rise to repolarization of the action potential* (Figure 12).

To examine the time course of Na⁺ entry and K⁺ exit in the axon during the generation of action potentials, load and run ACTN_POT.CCS with the O and B keys. This plot also illustrates the independent activation and inactivation processes of the Na⁺ current during action-potential generation at the bottom of the screen. Note that the degree of activation of the Na⁺ current (denoted as mNa on your computer screen) goes up to complete activation at the peak, although an increase in inactivation (denoted as hNa on your computer screen) also occurs (Figure 13).

The next section, "Neurophysiological Properties of Neurons," will extend these basic properties by demonstrating some of the wide variety of

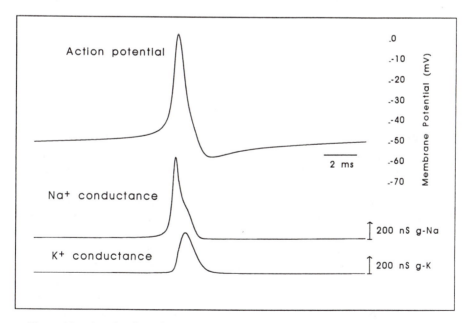

Figure 12. Amplitude and time course of the Na⁺ and K⁺ conductances (gNa, gK) during the generation of an action potential.

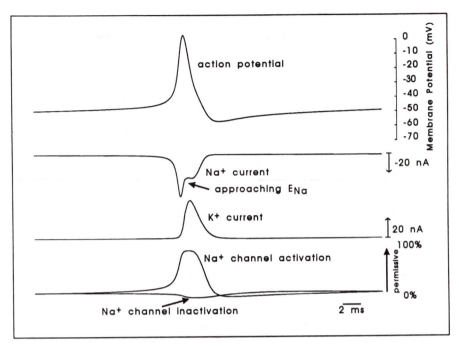

Figure 13. Amplitude and time course Na$^+$ and K$^+$ currents during generation of an action potential. Also illustrated is the degree of activation and inactivation of the underlying Na$^+$ channels. Upward represents increased activation and decreased inactivation.

ionic currents that are present in neurons of the central and peripheral nervous systems.

Study Questions: Mechanisms of Action-Potential Generation

1. Which ion is important for the upswing (rising portion) and which is important for the downswing (falling portion) of the action potential? In what direction do these ions move (outside the cell to in or inside to out)?
2. What is meant by the terms activation and deactivation? How about inactivation and deinactivation? How is this important in considering the Na$^+$ current underlying action-potential generation?
3. How do the inward Na$^+$ and outward K$^+$ currents compare in their rates of activation? (This can be examined in V-CLAMP with NA_K.VCS.) Which current, the Na$^+$ or the K$^+$, activates at more hyperpolarized levels? (This can be examined in V-CLAMP by single stepping through NA_K_IV.VCS.) How are these features important for action-potential generation?
4. Why doesn't the action potential peak at the equilibrium potential for

Na$^+$? (The answer to this can be found by closely examining the ionic currents by running ACTN_POT.CCS.)

5. What four factors determine the amplitude of an ionic current at any particular time? (Closely examine the Na$^+$ current in Figure 13.)

Answers are on page 70.

Neurophysiological Properties
of Neurons

You now have at least a working knowledge of action-potential generation in the squid giant axon. Early investigators of the mammalian brain imagined that the central nervous system (CNS) consisted of a complex interconnected network of neurons that possessed properties essentially identical to those determined by Hodgkin and Huxley for the squid giant axon (i.e., they generated simple action potentials though Na^+ entry and K^+ exit) and that communicated to one another through simple excitatory, and perhaps inhibitory, connections. Interestingly, the detailed investigation of neurons at all levels of the nervous system (including all levels of the animal kingdom and even some plants) revealed a complex set of ionic conductances that could be mixed together in various manners to give each different cell type a propensity to generate unique patterns of action potentials. Fortunately, once you understand how one ionic current works and how it influences the membrane potential of the cell, you then have the basic knowledge needed to understand complex neurons with their dozens of ionic currents! The understanding of the principles of more complex patterns of activity is merely the addition of currents that all follow the same basic rules. Let's now investigate just a few of these different neuronal ionic currents.

First, imagine that we have just read Hodgkin and Huxley's series of papers in the *Journal of Physiology* and are quite impressed, but at the same time are wondering if neurons generate action potentials in a similar manner, or if perhaps they have even more complex ionic currents. To perform similar experiments we need cells that can withstand being impaled by two glass microelectrodes (one for current injection and one for recording voltage) and in a preparation that does not move! This rules out any mammalian system, for the brain pulsates during the cardiac cycle and

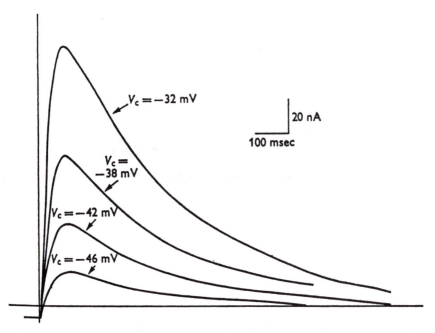

Figure 14. Examples of a transient outward current, termed the A-current, that is activated in neurons from nudibranchs by depolarization. Holding the cell at -100 mV in voltage clamp and stepping to the membrane potential shown resulted in the activation of an outward K^+ current that inactivated over a period of hundreds of milliseconds. Note that the current becomes larger with depolarization from -46 to -32 mV. (Adapted from Connor & Stevens, 1971a.)

respiration,[17] it is impossible to impale the small neurons of the brain with two microelectrodes, and no one has yet figured out a way to keep mammalian brain cells alive in vitro. Therefore, we turn to the relatively large neurons of marine gastropods, such as the invertebrate *Anisodora*. Using isolated neurons, and methods based upon those of Hodgkin and Huxley, researchers John Connor and Chuck Stevens (Connor & Stevens, 1971a) indeed found similar ionic currents to those described by Hodgkin and Huxley. However, they also found a new type of K^+ current that was not found in the squid giant axon. In the invertebrate neurons, depolarization of the membrane gave rise to the activation of a K^+ current that, unlike the K^+ current of Hodgkin and Huxley, inactivated (turned-off) with time despite the maintained depolarization (Figure 14). Incorporating this new "transient" K^+ current into a model of action-potential generation suggested that the current was active in-between action-potential generation and therefore acted to slow the rate of action-potential generation (Figure

---

[17]Pulsation occurs when an opening in the skull is made to gain access to the neural tissue below.

15C) so that the cell was capable of firing at slow firing rates, compared to the squid axon (Connor & Stevens, 1971b). This K^+ current was termed the "A-current." We shall now examine this current and its influence on the pattern of action-potential generation in our own experiments.

Experiment 11: Properties of a Transient K^+ Current (A-Current)

Let us consider the A-current by examining its effects on the firing pattern generated by the modeled cell (see also Chapter 5, in *Neurobiology*). From within C-CLAMP, load and execute IA.CCS. with the O (open) and B (begin) keys. You will see that the generation of the action potential is delayed (Figure 16), and a plot of the A-current reveals this current to increase and then decrease over time. In real experiments, the A-current can be blocked relatively selectively through the application of 4-amino-pyridine.[18] Here we can block the A-current by reducing its maximal conductance (gA) to 0.

After reducing gA from 1 to 0, press the Y (overlay) key and compare how the cell fires with and without the A-current (Figure 16). It would appear that the current activates during depolarization of the cell, and in so doing delays the onset of action-potential generation, but does not greatly influence the shape of individual action potential. Let us now switch the cell into the voltage-clamp mode of recording by quitting C-CLAMP and running V-CLAMP.[19] Now load the file IA.VCS with the O key and single step through it by pressing the I (individual steps) key repeatedly (Figure 17A). In this paradigm we note that depolarizing the cell results in the activation and then inactivation of an outward current and that increasing the level of depolarization results in the current becoming larger. Note that the increased amplitude is a product of both *increased activation of the current* and an *increase in the "driving force" on the K^+ ions*, since we are moving the membrane potential away from E_K. Now let's examine the ionic basis of the A-current by increasing the extracellular concentration of $[K^+]_o$. Change $[K^+]_o$= from 3.1 mM to 25 mM, and in so doing, change the equilibrium potential for K^+ from -100 to -60 mV (this would be achieved in vitro by changing bathing solutions while keeping the microelectrodes in the cell!). Again, single step through different voltages with the I key and note that the current is no longer always outward, but that the first currents to be activated are inward (Figure 17B), since the current now activates at a membrane potential negative to the reversal potential of K^+, and subsequently K^+ ions move *into* the cell, instead of out of the cell.

Now let's imagine that you have developed a preparation for the study

---

[18]Multiple sclerosis is a disease of blocked motor transmission, which thereby generates difficulties in movement. In some patients with multiple sclerosis, 4-aminopyridine partially alleviates this problem through increasing neuronal and axonal excitability!

[19]Modern intracellular amplifiers allow the switching from current-clamp to voltage-clamp recordings with the mere push of a button—a far cry from the days of build-it-yourself electrophysiology.

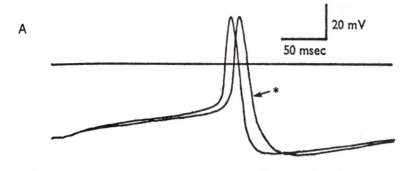

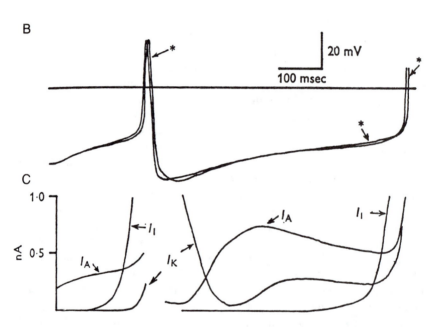

Figure 15. Computational model of action-potential generation in neurons that possess an A-current. *A* and *B*, Example of a real action potential and one computed (denoted by *) according to equations derived from the properties derived by voltage-clamp analysis of the underlying currents. *C*, Computed membrane currents associated with the generation of the action potential in *B*. I_I refers to the inward Na^+ current and I_K is the delayed rectifier K^+ current. Currents are not shown during the generation of the action potential, owing to their large amplitude. (Adapted from Connor & Stevens, 1971b).

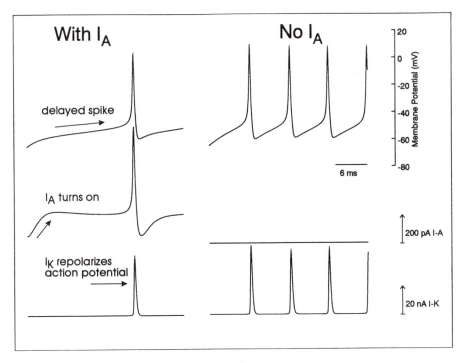

Figure 16. Activation of the transient K$^+$ current I$_A$ results in a delay in action-potential generation in the modeled neuron.

of vertebrate cells of a sympathetic ganglia from the bullfrog. Injection of current into these cells results in the activation of a complex pattern of action potentials that result from activation of a wide variety of currents. Eventually, you demonstrate that there are at least four different K$^+$ currents. In addition to the A-current, there are also two that are activated by the entry of Ca^{++} into the cell during action potentials (I$_C$ and I$_{AHP}$) and one that is activated by depolarization (I$_M$). At about the same time, other investigators are perfecting techniques to record intracellularly in mammalian neurons in vitro. These techniques include the culturing of mammalian CNS neurons and the in vitro slice technique[20] (see Chapter 7 in *Neurobiology*). Investigations of CNS neurons reveal that cortical and hippocampal pyramidal cells in animals ranging from rodents to humans also have currents similar to those of the bullfrog sympathetic ganglia, indicating that these currents have wide applicability. Let's consider these new K$^+$ currents now.

---

[20]The in vitro slice technique allows the experimenter to maintain a thin slice, approximately 0.5 mm thick, of brain tissue alive and healthy in vitro for several hours. It is widely used to study the physiology of the central nervous system.

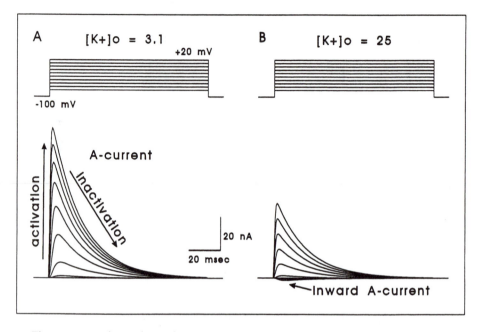

Figure 17. Voltage-clamp demonstration of the A-current in the modeled neuron. Depolarization of the neuron results in the activation of an outward K⁺ current that inactivates with time. Increasing the extracellular concentration of K⁺ results in a shift in the reversal potential; consequently, activation of the A-current at some steps results in an inward K⁺ current.

Experiment 12: I_L and I_C—High-Threshold Ca⁺⁺ Current and Ca⁺⁺-Activated K⁺ Current

In the squid giant axon the repolarization of the action potential results from inactivation of the Na⁺ current and the activation of a K⁺ current (the delayed rectifier), which is activated by the depolarization of the membrane during the action potential. However, in many different cell types it has been found that removal of extracellular Ca⁺⁺, or block of Ca⁺⁺ channels with nonpermeant divalent cations (such as cadmium [Cd⁺⁺]), results in a reduction of the repolarizing phase of the action potential and of the hyperpolarization of the membrane following the action potential (e.g., Figure 18B). This result suggests that Ca⁺⁺ enters into the cell during the action potential, and this then activates K⁺ currents that help to repolarize the action potential!

To model such a K⁺ current, we need not only a model of this current, but also a model of a Ca⁺⁺ current that allows the entry of Ca⁺⁺ into the cell during the generation of the action potential. Therefore, we will now add to our model two additional currents: a high-threshold (activates only at membrane potentials positive to approximately −30 mV) Ca⁺⁺ current,

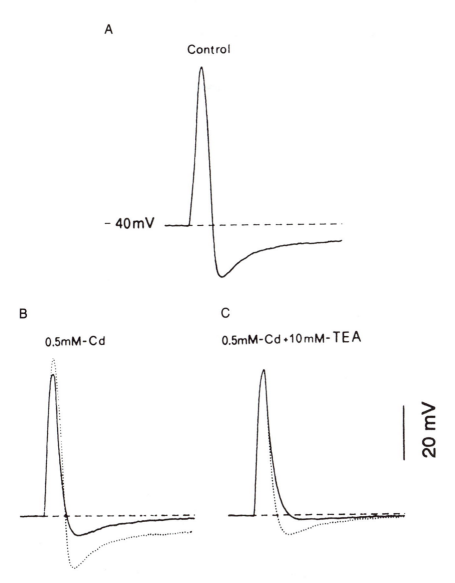

Figure 18. Importance of a Ca^{++}-activated K^+ conductance known as I_C in action-potential generation in mammalian cells. *A*, Intracellular recording of an action potential from a rat sympathetic neuron. *B*, Block of Ca^{++} entry into the cell through bath application of an ion that blocks Ca^{++} channels (Cd^{++}) results in a marked reduction of the hyperpolarization following the action potential (compare solid line with dashed line). *C*, Addition of tetraethylammonium (TEA) to block other K^+ conductances results in further reduction of the afterhyperpolarization. (Adapted from Belluzzi & Sacchi, 1991.)

39

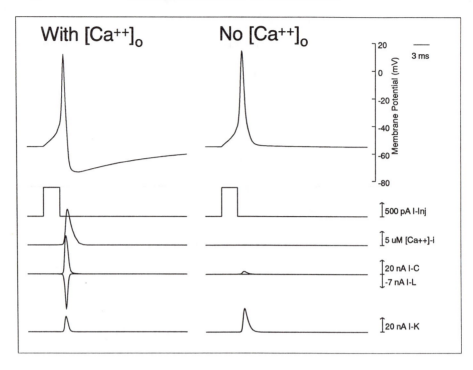

Figure 19. Addition of a Ca^{++}-activated K^+ current and a high-threshold Ca^{++} current helps to repolarize the membrane after each action potential in the modeled neuron. Reduction of I_C results in reduction of the hyperpolarization following the action potential, an increase in the duration of each action potential, and an increase in the amount of I_K activated.

termed I_L; and a Ca^{++} sensitive and voltage-sensitive K^+ current, termed I_C (see Chapter 5 in *Neurobiology*, Shepherd, 1994; Storm, 1990). Run C-CLAMP and then load and run the parameter file IC.CCS. Here we see that during the generation of an action potential, the intracellular level of Ca^{++} increases.[21] This increase in $[Ca^{++}]_i$ results in the activation of the outward K^+ current I_C, which then helps to repolarize the action potential (Figure 19). To test this hypothesis, move the cursor to $[Ca^{++}]_o=$ and reduce this to 0.01 mM and press the **Y** key. Notice now that I_C is only minimally activated during the action potential and that the hyperpolarization that follows the action potential is reduced in amplitude. However, the action potential does not broaden as much as we might have suspected, based upon the amplitude of I_C prior to its block. This is because the other K^+ current, I_K, becomes larger (owing to small increases in duration and amplitude of the action potential) and therefore compensates in part for the

---

[21]The Ca^{++} concentration is calculated as the concentration in the 100 nm of space just under the membrane with a rate of Ca^{++} buffering that is proportional to $[Ca^{++}]_i$.

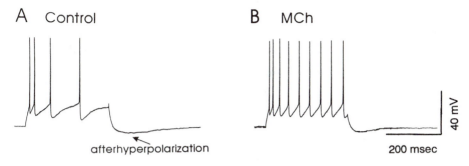

A Control   B MCh

40 mV

afterhyperpolarization   200 msec

Figure 20. Response of a human cortical pyramidal cell to membrane depolarization. *A*, Intracellular injection of a depolarizing current pulse results in a series of four action potentials, the rate of generation of which slows down with time, a process known as spike frequency adaptation. When the depolarizing current pulse ends, it is followed by a slow afterhyperpolarization that is mediated in part by a Ca^{++}-activated K^+ current known as I_{AHP}. *B*, Activation of cholinergic muscarinic receptors with MCh (methylcholine) results in a marked reduction of spike frequency adaptation and a reduction of the slow afterhyperpolarization. These effects occur through a reduction in I_{AHP}.

loss of I_C (Figure 19). This is an important point, for neurons are dynamic systems in which currents interact in a manner that is not easily understood without the aid of computer models such as this one.

Experiment 13: I_{AHP}—Slow, Ca^{++}-Activated K^+ Current: Regulator of Cell Excitability

Some neurons in the nervous system display yet another type of Ca^{++} activated K^+ current. For example, intracellular injection of a depolarizing current pulse into a cortical pyramidal cell from the human neocortex (a slice of brain tissue obtained during the neurosurgical treatment of epilepsy[22] and kept alive in a special chamber in vitro) results in a series of action potentials that are followed by a slow hyperpolarization known as an "afterhyperpolarization" (Figure 20A). This afterhyperpolarization is largely the result of the activation of a slow K^+ current, known as I_{AHP}, that is activated by the entry of Ca^{++} during the series of action potentials (see Chapter 5 in *Neurobiology*). As each action potential occurs, Ca^{++} enters the cell through high-threshold Ca^{++} channels. The increases in intracellular Ca^{++} activate I_{AHP}, which then hyperpolarizes the cell slightly, resulting in a decrease in rate of action-potential discharge, a process known as spike frequency adaptation (Figure 20A). Activation of a variety

---

[22]Some epileptic patients suffer from reoccurring seizures that are not alleviated through pharmacological treatment and that occur in response to abnormal discharges in a well-localized portion of damaged or improperly functioning cortex known as the "focus." Removal of this focus in these patients through neurosurgery very often rids the patient of seizures.

of receptors, including acetylcholine muscarinic receptors, on cortical pyramidal cells reduces I_{AHP} and therefore increase the rate of action-potential discharge (Figure 20B; reviewed in McCormick, 1992; Nicoll et al., 1990). This increase in excitability in cortical pyramidal cells has been proposed to be an important factor in keeping the cerebral cortex responsive during waking and attentiveness and during dream sleep and less responsive during drowsiness and inattentiveness and during nondream sleep (slow-wave sleep; see McCormick, 1992; Steriade & McCarley, 1990).

To examine I_{AHP} in C-CLAMP, load and run IAHP.CCS with the O and B keys. Note that during the generation of a train of action potentials, the frequency of firing slows down and there is an afterhyperpolarization following the train of action potentials (Figure 21). Examining Ca^{++} entry into the cell and the amplitude–time course of I_{AHP} with the computer model suggests that the increases in intracellular Ca^{++} give rise to the activation of this current (Figure 21). To mimic the block of I_{AHP} on a human cortical pyramidal cell by acetylcholine, move the cursor to $gAHP=$ and reduce this to 0 and press the Y key. Note that now the cell fires a more rapid train of action potentials that do not show signs of spike frequency adaptation (Figure 21). You can also test the Ca^{++} dependence of this current by reducing $[Ca^{++}]_o$ to 0.1 mM with g_{AHP} returned to 0.06 nS and examining the effects.

Experiment 14: Sleep and Waking in Single Neurons: I_T—Transient and Low-Threshold Ca^{++} Current

The first investigators to record from neurons in the brains of sleeping animals made a remarkable discovery. Instead of being silent during nondreaming sleep (slow-wave sleep) as one might expect, Herbert Jasper, David Hubel, and Edward Evarts independently found that many cells were discharging in bursts of action potentials instead of the more independent series of single spikes typical of the waking animal (e.g., Jasper, Rici, & Doane, 1958). Indeed, intracellular recordings during slow-wave sleep from the thalamic relay cells that transmit information from the retina to the visual cortex revealed the presence of slow spikes underlying the burst discharges of the sleeping brain (Figure 22; bursts). The transition to waking or dreaming sleep was associated with depolarization of the membrane potential, a lack of these slow spikes and burst discharges, and the generation of action potentials in a more regular manner (dreaming sleep is also known as REM sleep for the rapid eye movements associated with the eyes darting back and forth during dreams; Figure 22). Additional intracellular investigations, both in vivo and in vitro by Mircea Steriade, Martin Deschênes and colleagues, and Henrik Jahnsen and Rodolfo Llinás revealed the ionic mechanisms of this striking and important change in neuronal activity in the transition from sleep to waking (see Chapter 25 in *Neurobiology*, Steriade & Deschénes, 1984; Jahnsen & Llinás, 1984a, 1984b).

Intracellular recordings from thalamic relay cells reveal that they have two modes of action-potential generation. At relatively hyperpolarized membrane potentials, intracellular injection of a depolarizing current pulse results in the activation of a low threshold Ca^{++} spike that then activates a burst of 3 to 5 action potentials (Figure 23; burst mode). However, if the cell is at -63 mV, intracellular injection of the same current pulse now

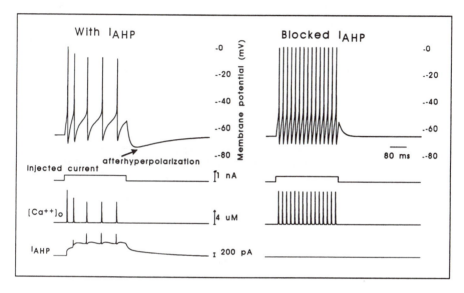

Figure 21. The slow Ca^{++}-activated K$^+$ current I_{AHP} results in slowing down of rate of action-potential generation and a slow hyperpolarization after a train of action potentials in the modeled neuron.

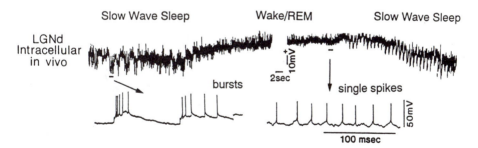

Figure 22. Intracellular recording from a thalamic relay cell in the transition from slow-wave sleep to rapid eye movement (REM), or dreaming, sleep. During slow-wave sleep, this cell generates bursts of action potentials (labeled bursts and expanded for detail). In the transition to REM sleep (or waking) the membrane potential depolarizes by 10 to 20 mV and the cell generates series of single action potentials (labeled single spikes and expanded for detail). The transition back to slow-wave sleep is again associated with hyperpolarization of the membrane. (Adapted from Hirsch et al., 1983).

results in only the passive response of the cell which is shaped by the resistive and capacitive properties of the cell! If the cell is further depolarized to −53 mV, then the same current pulse results in a train of action potentials that, unlike those in cortical pyramidal cells, *does not* exhibit marked slowing of the rate of action-potential discharge (Figure 23; transfer mode). In this manner, even single neurons exhibit changes in activity in relation to the sleep and wake cycle. The presence in thalamic cells of an extra mode of action potential generation, the burst mode, during slow-wave sleep is due to the properties of a special type of Ca^{++} current known as the low threshold, or transient, Ca^{++} current.

To examine the properties of this current, load and run IT.CCS in C-CLAMP with the O and B keys (Figure 24A). Note that depolarization of the model cell results in the generation of a burst of action potentials riding on top of a slow Ca^{++} spike. The amplitude and time course of the current underlying this slow spike, the T-current, can also be seen on the computer screen (Figure 24A; T-current). Now, depolarize the neuron to −60 mV by changing *Base current (nA)=* to **0.18** and *Starting Vm (mV)=* to **−60** and press **B**. Now the cell generates a steady train of action potentials and no burst discharges (Figure 24B)! The lack of burst firing at depolarized membrane potentials indicates that the current underlying the burst is highly voltage dependent and is *inactivated* by depolarization. To examine this a bit more, let's isolate the slow spike from action-potential generation by applying tetrodotoxin (TTX) to the cell. Load IT.CCS and move the

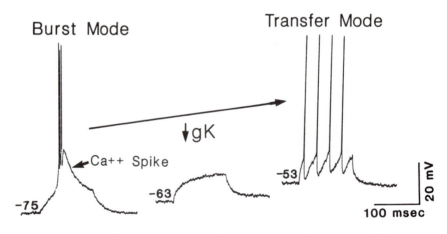

Figure 23. Two different firing modes depend upon membrane potential in single thalamic relay neurons. Intracellular injection of a depolarizing current pulse into a thalamic relay neuron maintained in a slice in vitro results in a slow Ca^{++} spike and burst of action potentials if the cell is at −75 mV, a passive response if the cell is at −63 mV, and a series of four individual action potentials if the membrane is tonically depolarized to −53 mV. The burst discharge at −75 mV is due to the presence of a strong low-threshold Ca^{++} current in these cells. (For more information see Jahnsen & Llinás, 1984a, 1984b; Huguenard & McCormick, 1992.)

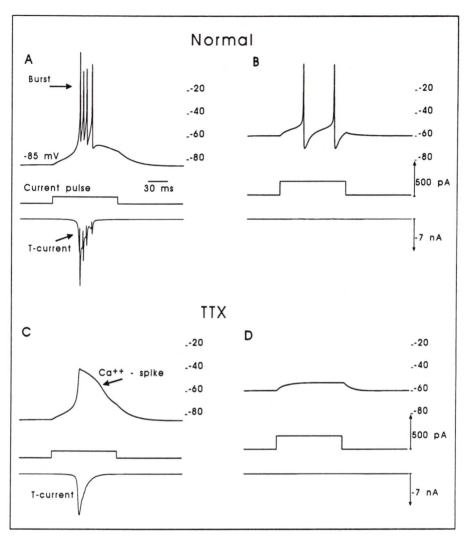

Figure 24. The transient Ca^{++} current generates burst discharges at hyperpolarized membrane potentials in the model cell. *A*, Injection of a depolarizing current pulse into this model "thalamic relay" cell results in a burst of action potentials. The burst is due to the activation of I_T. *B*, Depolarization of the cell to -60 mV inactivates the T-current and now the cell does not burst. *C* and *D*, Block of the Na^+ current with tetrodotoxin (TTX) reveals the underlying slow Ca^{++} spike and the lack of this Ca^{++} current when the cell is tonically depolarized to -60 mV.

cursor to $gNa=15$ and change this to 0, thereby blocking the voltage-dependent Na^+ conductance. Now press **B** to repeat the current pulse and note the slow spike on top of the passive membrane response (Figure 24C). The kinetics of the current underlying this slow spike are substantially slower than those underlying the fast action potential, thereby giving rise to this prolonged depolarization of the cell. Now you experiment with the ion concentrations in the bathing medium and find that when you reduce $[Ca^{++}]_o$ to 0.1 mM, you block the slow spike. This indicates that the slow spike is generated by the entry of Ca^{++} into the cell. In fact, this spike is generated by a Ca^{++} current known as the *transient current* (T-current). Like the Na^+ current underlying action-potential generation, the T-current inactivates with depolarization; therefore, steady depolarization of the cell results in complete inactivation of this current and a loss of these slow spikes. This property of the T-current underlies the ability of thalamic neurons to change from a pattern of burst firing during sleep to one of normal action-potential generation during wakefulness. How is this depolarization naturally achieved? Thalamic neurons are depolarized by the release of neurotransmitters from the brain stem systems that are responsible for keeping us awake during the day (see Chapter 25 in *Neurobiology*). The release of these neurotransmitters, such as acetylcholine, reduces the resting conductance of the membrane to K^+ ions, thereby resulting in a maintained depolarization of thalamic cells during wakefulness (see McCormick, 1992; Chapter 25 in *Neurobiology*). Interestingly, some drugs that have sedative side effects, such as antihistamines, block the receptors involved in maintaining this depolarization of central neurons. For example, activation of the H_1 subtype of histaminergic receptors on thalamic relay cells reduces *pKleak*, resulting in a tonic depolarization of these cells. The active ingredient in over-the-counter sleeping pills is an H_1-receptor antagonist and, by blocking the depolarizing actions of histamine, hyperpolarizes your thalamic neurons back into the sleep mode (McCormick & Williamson, 1991)!

Experiment 15: I_M—Depolarization and Slowly Activating K^+ Current

So far we have seen that neurons can generate a variety of different patterns of action potentials through interaction of different ionic currents and that these patterns can be altered by neurotransmitters through changes in these currents. The last example of this type of neuromodulation that we will examine was discovered in the sympathetic ganglion cells of the bullfrog by David Brown and Paul Adams (see Chapter 18 in *Neurobiology*). These investigators recorded from sympathetic ganglion cells in an attempt to reveal the mechanisms by which the transmitter released by the preganglionic neurons in the spinal cord (acetylcholine) modulates the excitability of the ganglionic cells through the activation of muscarinic receptors. Interestingly, they found that the amplitude of the slow excitation (depolarization) resulting from the activation of muscarinic receptors by the release of acetylcholine is very dependent upon the membrane potential of the cell, becoming larger with more depolarized membrane potentials. Brown and Adams demonstrated that this marked voltage dependence of the slow cholinergic excitation was due to the reduction by acetylcholine of a highly voltage-dependent K^+ current, which they termed the "M-current," owing to its block by activation of muscarinic receptors (Figure 25).The M-current is a K^+ current that upon depolarization of the cell is activated over a period of tens of milliseconds. For example, if we were to voltage clamp the cell and step from, say −65 to −45 mV, the M-current would slowly turn on (Figure 25A). Stimulation of the preganglionic nerve to release acetylcholine results in a marked suppression of this slowly activating K^+

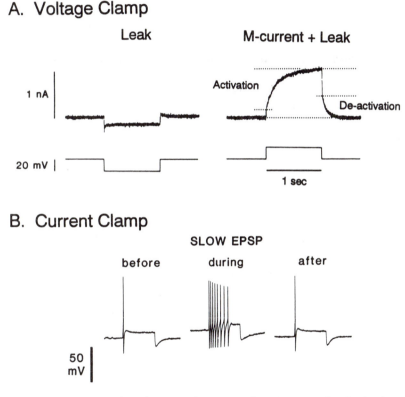

Figure 25. Properties of a K^+ current known as the M-current that is slowly activated by depolarization. *A*, In voltage-clamp mode, hyperpolarization of the cell by 20 mV reveals only the leak membrane conductance (trace labeled *Leak*). In contrast, depolarization of the cell by 20 mV results in the activation of a slow outward current (traces labeled *M-current + Leak*). Stepping back to the resting level results in turning off the M-current (labeled *De-activation*). *B*, Intracellular injection of a depolarizing current pulse during current-clamp recording before, during, and after recovery from the release of acetylcholine (slow EPSP). Note the marked increase in the number of action potentials associated with block of the M-current by acetylcholine. (Adapted from Jones & Adams, 1988.)

current. The activation of the M-current by depolarization results in a reduction of cell excitability such that the depolarizing current pulse may generate only one action potential (Figure 25B; before and after recovery from nerve stimulation). However, if the M-current is reduced by release of acetylcholine, then the same depolarizing current pulse will generate a series of action potentials (Figure 25B; during the effect of acetylcholine).

Load and run IM.CCS in C-CLAMP for an example of the manner in which these cells fire prior to application of acetylcholine (Figure 26A). You can now mimic the effect of release of acetylcholine by reducing gM= to 0 and press the Y key. Note that the excitability of the cell has greatly increased such that the neuron now generates multiple action potentials, where it used to generate only one (Figure 26B). First you suspect that this may be due to Ca^{++}-activated K^+ conductances, but you notice (after reloading IM.CCS) that reducing $[Ca^{++}]_o$ to 0.1 mM does not block this property. In an effort to understand what current underlies this behavior of the neuron, you put the cell into voltage-clamp mode by quitting C-CLAMP and running V-CLAMP and apply a depolarizing voltage step by loading and running IM.VCS. Here you find an outward current that is slowly turned on (activated) by the depolarization (Figure 27). Again, by changing extracellular K^+ concentrations you demonstrate that this is a K^+ current (you can try this now). You then apply acetylcholine to the cell and

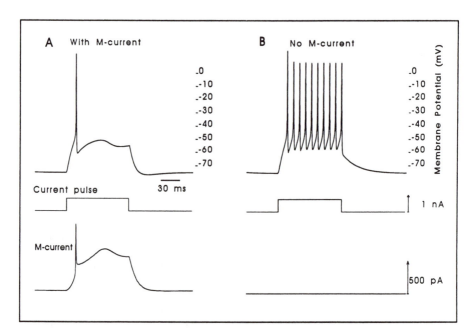

Figure 26. Depolarization activates a K^+ current known as the M-current, which subsequently prevents the generation of a train of action potentials in the model cell (A). Block of the M-current restores the ability of the cell to tonically fire (B).

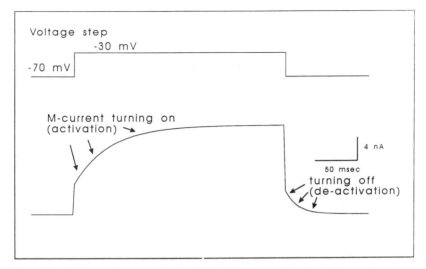

Figure 27. Voltage-clamp analysis of the M-current. Stepping the membrane potential of the model cell from -70 to -30 mV results in the slow activation of an outward K^+ current (M-current). Returning the membrane potential back to -70 mV turns the M-current off (a process known as deactivation).

note that this K^+ current is blocked by the activation of muscarinic receptors, which prompts you to dub it the "M-current" (see Brown, 1988; Chapter 18 in *Neurobiology*).

Congratulations! You have successfully made it through about 40 years of research and should now have an appreciation for some of the basic patterns of activity that neurons can generate intrinsically. However, neurons are only the building blocks of neural circuits, and to build these neural circuits, you need to connect cells through synaptic connections. So let's consider these now.

Study Questions: Multiple Ionic Currents in Central Neurons

1. What is a major difference between the A-current and the "delayed rectifier" K^+ current that repolarizes action potentials? (Compare Figures 10 and 17.)
2. How are Ca^{++}-activated K^+ currents important for action potential generation? (See Figures 19 and 21.)
3. Which current allows cells to generate bursts of action potentials at one membrane potential and trains of action potentials at another (Figure 24)? Why?
4. How is the M-current different from the "delayed rectifier" K^+ current underlying action-potential repolarization (compare Figures 27 and 10). What effect does the M-current have on the response of the cell to depolarization?

5. By blocking which ionic currents can neurotransmitters such as acetylcholine increase the excitability and responsiveness of neurons?

Answers are on page 70–71.

Synaptic Potentials

It was once believed that synaptic connections in the nervous system may be all of the excitatory type. However, the classic studies of spinal reflexes by Charles Sherrington (1906) demonstrated that central inhibitory, as well as excitatory, mechanisms were present. When Sir John Eccles recorded intracellularly from spinal cord motor neurons in the 1940s and 1950s, he found that not only were excitatory postsynaptic potentials (EPSPs) a feature of local circuit activity, but that inhibitory postsynaptic potentials (IPSPs) were also prominent (see Eccles, 1957). We now know that in considering fast (millisecond) neurotransmission, there are two basic types: excitatory and inhibitory. The difference between these is that excitatory neurotransmission *increases the probability* of generating an action potential, while inhibitory transmissions *decreases* it. The neurotransmitter most widely used in the brain for excitatory neurotransmission is believed to be glutamate or some other excitatory amino acid that can activate both postsynaptic AMPA/kainate and NMDA receptors (see Chapters 7 and 8 in *Neurobiology*). The most prevalent inhibitory neurotransmitter in the nervous system is GABA, which can activate both $GABA_A$ and $GABA_B$ receptors. Let's consider each of these now.

Experiment 16: Excitatory Postsynaptic Potentials

Let's say you are recording from a human or rodent hippocampal or cortical pyramidal cell (or a neuron just about anywhere in the brain of just about any type of animal, for that matter) and activate an excitatory afferent pathway through the delivery of a brief electrical stimulus to the axons of that pathway. What you record is a rapid EPSP followed by a biphasic hyperpolarization due to IPSPs (Figure 28; right-hand trace). You suspect that the inhibitory postsynaptic potentials may be coming from the discharge of local GABAergic neurons, and indeed, when you record from one of these you find that these cells discharge repetitively in response to the brief afferent stimulation (Figure 28; left-hand trace).

To examine the effects of this sequence of postsynaptic potentials on the neuron, load and run EP_IPSP.CCS with C-CLAMP. The excitatory pathway is due to release of glutamate onto the neuron by the afferents you stimulated, while the following inhibitory potentials are due to the release of GABA by local GABAergic interneurons. These inhibitory cells were also excited by the release of glutamate by the afferents you stimulated (Figure 29). Now you can isolate the EPSP from the biphasic IPSP by applying antagonists to GABA receptors. In the modeling program, move the cursor to *gIPSP*= and type in 0. Now press **Y** (overlay) to compare the EPSP before and after block of inhibition. Here, notice that the EPSP is substantially larger after the block of inhibition (Figure 29). In real neuronal circuits, this strong increase in the amplitude of EPSPs after the block of inhibition can result in "runaway" excitation between interconnected excitatory neurons, and therefore an epileptic seizure. Controlling the balance between excitation and inhibition is a major goal in the pharmacological treatment of epilepsy.

Through investigating the pharmacology of the isolated EPSP, you discover that it is mediated by two different types of receptors that are selectively activated by the agonists AMPA and NMDA. By investigating the voltage dependence and ion sensitivities of AMPA- and NMDA-receptor–mediated responses, you find that both of these receptors/channels pass

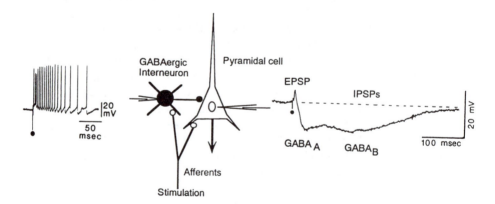

Figure 28. Activation of excitatory afferents to human cortical pyramidal cells results in excitatory followed by inhibitory postsynaptic potentials. Electrical stimulation of excitatory afferents (*dots*) results in a strong discharge of action potentials in local intracortical inhibitory (GABAergic) interneurons (*left*) as well as directly depolarizing pyramidal neurons (*right*, EPSP). However, the strong discharge in the GABAergic neurons results in hyperpolarizing (inhibition) of the pyramidal cell through increases in a Cl⁻ conductance through GABA$_A$ receptors and increases in a K⁺ conductance through GABA$_B$ receptors.

Na$^+$ and K$^+$ ions and therefore generate responses that reverse around 0
mV (see Figure 1). You can examine this by changing *Base current (nA)=*
to **1.78** and changing *Starting Vm (mV)=* to **20**. Now press Y and note
that the EPSP has reversed to hyperpolarizing, owing to the membrane
potential being positive to the reversal potential (Figure 29; reversed EPSP).

The NMDA portion of the EPSP has an unusual voltage dependence: it
becomes larger with depolarization, even though you are closer to the reversal potential (see Chapter 7 in *Neurobiology*)! To examine this, we isolate the NMDA component by blocking the AMPA component with a spe-
cific antagonist. Load and run the file NMDA.CCS to do this (Figure 30).
Now activation of the afferents generates an NMDA-receptor—mediated
EPSP only. Hyperpolarize the neuron by changing *Base current (nA)=* from
0 to **−0.52** nA, the *Starting Vm=* to **−90** and press Y. Note that the
NMDA-receptor–mediated EPSP is very small at this hyperpolarized
membrane potential (Figure 30). Change *Base current (nA)=* to **0.525** and
the *Starting Vm=* to **−30**, press Y, and note that the EPSP is larger at this
depolarized membrane potential, even though we are closer to the reversal
potential. By changing the extracellular concentration of different ions,
Linda Nowak, Philip Ascher, and colleagues realized that [Mg^{++}]$_o$ is important for this unusual voltage dependence of the NMDA-receptor-
mediated EPSP (Nowak et al., 1984). For example, change *Base current
(nA)=* to **−0.52** and the *Starting Vm=* to **−90** and press B (begin). Now

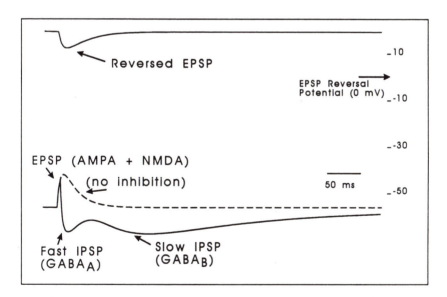

Figure 29. Excitatory and inhibitory postsynaptic potentials. Excitatory postsynaptic potentials are mediated by AMPA and NMDA receptors, while inhibitory
postsynaptic potentials are mediated by GABA$_A$ and GABA$_B$ receptors in the model
cell. Block of inhibitory potentials results in an increase in the amplitude of the
EPSP. The EPSP reverses polarity at 0 mV.

move the cursor to $[Mg^{++}]_o=$ and change this value from 1.2 to 0.01 mM and press **Y**. Note that the EPSP is much larger now after the removal of Mg^{++} from the bathing medium (Figure 30). To examine if the same is true at depolarized membrane potentials, change *Base current (nA)=* to 0.525 and *Starting Vm=* to -30 and press **B**. Move the cursor to $[Mg^{++}]_o=$ 0 and change this back to **1.2** and press **Y**. Note that there is a smaller effect of Mg^{++} ions on the NMDA-mediated EPSP at depolarized, versus hyperpolarized, membrane potentials (Figure 30).

To examine the voltage dependence of NMDA-receptor–mediated responses in more detail we need to examine these under voltage-clamp conditions. Quit the C-CLAMP program (Press **Q**), start V-CLAMP, and load and run NMDA.VCS. Here we have activated the NMDA-receptor–mediated synapses while voltage clamping the postsynaptic neuron to different membrane potentials (Figure 31A). By measuring the peak current

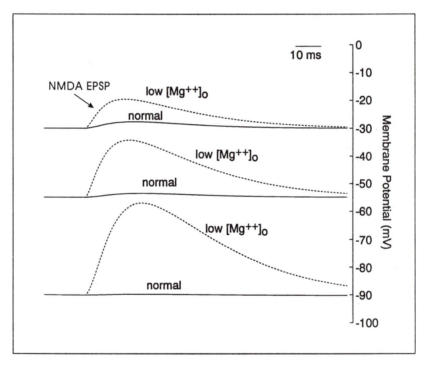

Figure 30. Unusual voltage dependence of NMDA-receptor–mediated component is due to voltage-dependent Mg^{++} block. When $[Mg^{++}]_o$ is reduced, the amplitude of the NMDA-receptor–mediated EPSP is larger at hyperpolarized membrane potentials (-90 mV) than at depolarized membrane potentials (-25 mV) as expected, owing to the approach to the equilibrium potential for NMDA currents (0 mV). In contrast, with Mg^{++} present in the extracellular medium (labeled normal), this ion suppresses the NMDA current at hyperpolarized membrane potentials (e.g., -55 and -90 mV) and therefore results in an unusual voltage dependence.

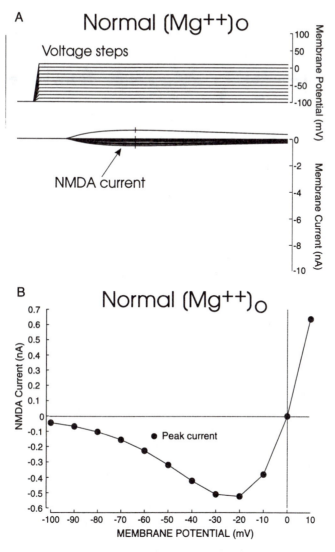

Figure 31. Voltage-clamp analysis of the voltage dependence of NMDA-receptor mediated–currents. *A*, Activation of the NMDA-receptor–mediated EPSP while voltage clamping the cell to various membrane potentials reveals the responses to become larger and then smaller as the membrane is depolarized. *B*, Plot of the peak NMDA current (denoted by + in *A*) versus membrane potential. Note that the NMDA current becomes larger as the membrane is depolarized from −100 to −20 mV and then becomes smaller and reverses polarity at 0 mV. (*Figure continues on following page*)

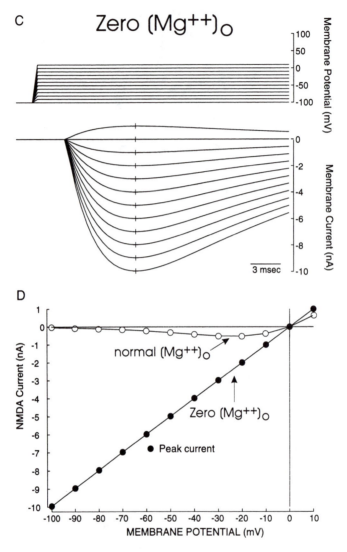

Figure 31. *Continued C*, Removing Mg^{++} from outside the cell results in linearization of the NMDA-receptor–mediated EPSP. *D*, Plot of the peak NMDA current versus membrane potential reveals that the amplitude of the EPSP is now determined by the distance to the reversal potential (0 mV) only. Note the difference in the NMDA current scale between *B* and *D*. The response with normal [Mg^{++}]$_o$ is also shown in *D* for comparison.

flowing through the NMDA channels at different membrane potentials, we get a plot of the voltage dependence of the NMDA current (Figure 31*B*). You can obtain this plot on your computer screen by pressing the Return key at the end of the simulation. Note that as the membrane potential depolarizes from -100 to -30mV, the NMDA current become larger, although at more depolarized levels it become smaller again and even reverses polarity at 0 mV (Figure 31*B*). If we reduce $[Mg^{++}]_o$ to 0.001 mM (do this now) and run the experiment again (press Return after the last voltage step), we find that the NMDA-receptor–mediated EPSP is no longer voltage dependent (Figure 31*C* and *D*)!

These results lead you to hypothesize that NMDA receptors demonstrate a marked voltage dependence because at hyperpolarized membrane potentials, the negative potential inside the cell attracts Mg^{++} to attempt to enter the cell. Once the Mg^{++} ions enter the NMDA channels they become "stuck" and therefore block conduction of ions through these pores. However, at depolarized membrane potentials, the attraction Mg^{++} to enter the cell is much less; therefore, when NMDA channels are open, only a few become "plugged" by Mg^{++} ions (see Mayer et al., 1984). If you like, you can perform similar experiments with the AMPA-receptor–mediated EPSP in isolation by loading AMPA.CCS or AMPA.VCS. You will find that AMPA receptors are *not* voltage dependent and are *not* affected by $[Mg^{++}]_o$.

Experiment 17: Inhibitory Postsynaptic Potentials

By blocking excitatory postsynaptic receptors pharmacologically and directly stimulating inhibitory interneurons, you can examine the inhibitory postsynaptic potentials in isolation from excitatory postsynaptic potentials. In C-CLAMP, load and run IPSPS.CCS. Note that again we have a biphasic, or two-part, inhibitory potential. Through pharmacological investigation, you determine that the first part is mediated through the activation of $GABA_A$ receptors, while the second part is mediated through $GABA_B$ receptors (Figure 32). Then you investigate the ionic mechanisms of the generation of these different IPSPs by changing the membrane potential and the extracellular concentration of ions in the bathing medium. Move the cursor to *Base current (nA)=* and change this from 0 to -0.38 nA and change the *Starting Vm=* to -85 mV and press Y. Notice now that the first IPSP is reversed (depolarizing), while the later IPSP is not (it is still hyperpolarizing) (Figure 32). This indicates that they are mediated by different ions. Let's examine the dependence of the different IPSPs on extracellular ion concentrations. First load and run IPSPS.CCS. You hypothesize that the early, $GABA_A$ IPSP is mediated by Cl^-, since your investigations of voltage dependence show that it reverses at the equilibrium potential for Cl^- (E_{Cl}; -75 mV). Therefore, you decrease $[Cl^-]_o$ from 120 to 7 mM (do this now) and press Y. Note that the $GABA_A$ IPSP is now depolarizing, indicating that changing the equilibrium potential for Cl^- so that it is positive to the

membrane potential of the cell changed the direction of Cl⁻ flow. Whereas
Cl⁻ originally flowed from outside the cell to in, it now moves from inside
the cell to out (Figure 32; $[Cl^-]_i=[Cl^-]_o$). Similarly, you change the extracellular concentration of K⁺ (first reload and rerun IPSPS.CCS) from 3.1
to 25 mM and *Base current (nA)=* to −0.5 and find that now the late,
GABA$_B$ IPSP is depolarizing, indicating that it is mediated by an increase
in K⁺ conductance (Figure 32; $[K^+]_o=25$).

An often confused aspect of synaptic transmission is the equating of depolarizing potentials with excitatory synaptic transmission and hyperpolarizing synaptic potentials with inhibitory synaptic transmission. However,
we have already seen that by hyperpolarizing the cell below E$_{Cl}$, a hyperpolarizing IPSP can become depolarizing. Does this make the previously
inhibitory synaptic potential excitatory? No. The reason is that even though
the IPSP is depolarizing, its equilibrium potential is still −75 mV and therefore 20 mV below the threshold for generation of an action potential (typically −55 mV). To illustrate this, load IP_EPSP.CCS and press B. This is
an isolated EPSP activated at −85 mV that makes the cell fire an action
potential (Figure 33; EPSP alone). Now move the cursor to *gEPSP=* and
change this from 0.15 to 0, thus turning off the EPSP, and move the cursor
to *gIPSP=* and change this to 0.2, thus turning on the IPSP (GABA$_A$ only
in this case). Now press **B** again and notice that the IPSP is depolarizing
(Figure 33; reversed IPSP). Now move the cursor to *gEPSP=* and change
this back to 0.15 nS, press **Y**, and notice that now the EPSP does not
generate an action potential (Figure 33; EPSP + IPSP), since the IPSP
"pulls" the peak of the EPSP towards E$_{Cl}$ (−75 mV), and therefore away

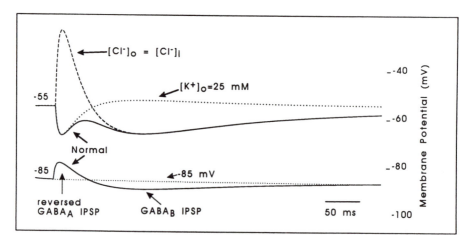

Figure 32. Ionic dependence of the two phases of IPSPs (GABA$_A$ and GABA$_B$).
Changing chloride concentration affects the first, GABA$_A$-mediated IPSP; while
changing potassium concentrations affects the later, GABA$_B$-mediated IPSP. The reversal potential of the GABA$_A$-mediated IPSP in normal solution is −75 mV, while
the reversal potential the GABA$_B$-mediated IPSP is −100 mV.

from action-potential threshold. Therefore, postsynaptic potentials that result from an increase in membrane conductance and that have a reversal potential below action-potential threshold (e.g., -55 mV) are inhibitory, even if they are depolarizing (see Chapter 7 in *Neurobiology*).

Summary—Building a Neural Network

The complex variety of ionic currents, of which we have just reviewed but a few, allows neurons in neural networks to fire in unique ways to facilitate their own particular role in neuronal processing. The connection of these neurons together with inhibitory, excitatory, and modulatory synaptic contacts allows for the generation of an even richer variety of patterned activity that may be useful for the coordinated performance of a motor task, the analysis of a visual scene, or the filtering of sensory information during sleep. We hope that the present computational model has brought the understanding of the neurophysiology of neurons just a little bit closer, so that some of you may go on to explain the mechanisms by which the nervous system carries out it's many varied functions.

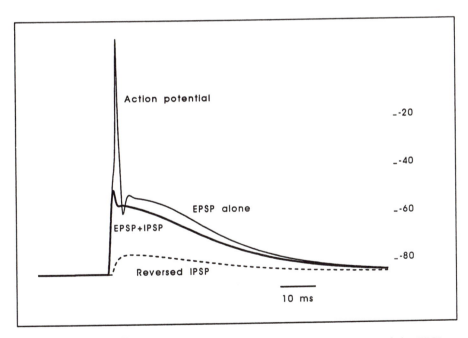

Figure 33. Reversed (depolarizing) IPSPs are inhibitory. Activation of the EPSP alone results in an action potential. Activation of the Cl^--mediated IPSP alone at a membrane potential of -85 mV results in a depolarizing IPSP (*dashed line*). Activation of the EPSP and IPSP together results in inhibition of the EPSP so that it no longer activates an action potential.

Study Questions: Synaptic Potentials

1. What is the effect of release of glutamate onto a cell? How about GABA? (See Figure 29.)
2. If the extracellular concentration of Cl^- was equal to the intracellular concentration, then would activation of $GABA_A$ receptors be inhibitory or excitatory? Why? (This can be tested by running IPSPS.CCS in C-CLAMP, making $[Cl^-]_o = [Cl^-]_i$, turning on the Na^+ and delayed rectifier K^+ currents to 10 and 2 μS, respectively.)
3. What are the three main factors controlling the amplitude of the NMDA-receptor–mediated EPSP?
4. Why do NMDA- and AMPA-receptor–mediated EPSPs reverse polarity at 0 mV?

Answers are on page 71.

Appendix A: Nernst and Goldman-Hodgkin-Katz Constant Field Equations

In this appendix we will briefly consider the Nernst and Goldman-Hodgkin-Katz (GHK) constant field equations. If you have two solutions with different concentrations of an ion and that are separated by a membrane that is permeable to that ion, then the ion will tend to move down its concentration gradient (through simple probability). However, movements of ions will also set up an electrical difference between the two solutions. This voltage will also affect the movement of the ion. The voltage difference across the membrane that is needed to just offset the tendency for the ion to move down its concentration gradient is the equilibrium potential and can be calculated using the Nernst equation.

$$E_X = RT/zF \cdot \ln([X]_o/[X]_i)$$

where R is the gas constant of 8.314 volts $\cdot$ coulombs/(T $\cdot$ mol), F is Faraday's constant 9.648×10^4 coulombs/mol, T is absolute temperature where T = 273.16 + temperature in Celsius, and z is the number and polarity of charges for each ion (e.g., +2 for divalent cations).

If the membrane is permeable to more than one ion, the membrane potential across the cell will be determined by the permeability of the membrane to each different type of ion and the concentrations of those ions. It is important at this point to differentiate between permeability and conductance. Permeability is the ability of a membrane to pass or flux ions. Conductance depends on both permeability and the presence of *permeant* ions. For example, a membrane that is permeable to K^+ would have very low conductance in the absence of K^+ ions. If the membrane is permeable to more than one ion, and if we know the relative permeabilities of the ions, then the resting membrane potential can be calculated by the GHK constant field equation.

$$E = RT/zF \cdot \ln\left((P_K[K]_o + P_{Na}[Na]_o)/(P_K[K]_i + P_{Na}[Na]_i)\right)$$

Here you can see that E depends not only on the concentrations of K^+ and Na^+ inside and outside the cell, but also on the relative permeabilities of these two ions, P_K and P_N: These are defined as $u\beta RT/aF$, where u is the mobility of the ion in the membrane; β is the partition coefficient between the membrane and the aqueous solution; a is the thickness of the membrane; and R, T, F are as above. In most neuronal membranes P_{Na} tends to be much lower than P_K, about 2 to 6%, and thus the resting membrane potential is much closer to the Nernst E_K than to E_{Na}. For a more detailed discussion and derivation of these equations, see Hille, (1984).

When the concentrations gradient of an ion is very large, the behavior of the conduction of that ion across the membrane deviates significantly from that predicted by Ohm's law and is more accurately described by the GHK constant field equation. Since the internal concentration of Ca^{++} is typically very low (e.g., 50 nM at rest), we chose to use GHK permeabilities to describe the two Ca^{++} currents, I_T and I_L, in the present model. In addition, the two leak currents, I_{Kleak} and I_{Naleak}, were also described by GHK permeabilities, since this more accurately portrays the properties of the resting membrane, which is permeable to both K^+ and to a lesser extent Na^+.

Appendix B: A Brief Explanation of How the Model Works

The model is based upon the methods pioneered by Alan Hodgkin and Andrew Huxley (1952), which we will summarize here. First, our model is known as a single-compartmental model, meaning that, unlike real cells, all of the ionic currents occur in the same place, with no passive or active conduction down processes (e.g., no dendrites or axons) or diffusion away from the single compartment.[23] Although this is a gross simplification of real neurons, the electrophysiological behavior of single-compartmental models are remarkably similar to those of the modeled cells and computationally are much faster to implement and easier to understand.

To model an ionic current, you need to know the following features:

1. Reversal potential.
2. Voltage dependence of activation and inactivation.
3. Time dependence (kinetics) of activation and inactivation at different membrane potentials.
4. Maximal conductance present in the cell (i.e., if all of the channels were open, what the total conductance would be).

Let's start by considering a current that does not inactivate, to make things simpler. Examples of currents that do not inactivate (i.e., do not turn off during a continued presence of the membrane potential in their activation range) are the M-current and the delayed rectifier K^+ current I_K studied by Hodgkin and Huxley. Hodgkin and Huxley modeled I_K with the following set of assumptions:

$$I_K = n^4 \cdot gK \cdot (E - E_K), \tag{1}$$

---

[23]The exception is Ca^{++}, which is assumed to diffuse away from the inner surface of a fictional membrane.

where n is an activation variable (or gate), gK is the maximal conductance, E is the membrane potential, and E_K is the reversal potential for K^+ ions. (Remember from Ohm's law that $V = I \cdot R$ or $I = V \cdot G$, since $G = 1/R$. Here $I = I_K$, $G = gK$, and $V =$ the driving force on K^+ or $E - E_K$.) n is a time- and voltage-dependent variable that follows first-order reaction kinetics of the form:

$$1 - n \underset{\beta_n}{\overset{\alpha_n}{\rightleftharpoons}} n. \qquad (2)$$

In this manner, the "n" state is considered permissive (e.g., channels open or activated), while the "$1 - n$" state is considered nonpermissive (e.g., channels closed or deactivated) for current flow through the membrane. Since all channels must be either in the permissive or nonpermissive states, the proportion of channels in either of the two states must add to 1 (e.g., $1 - n + n = 1$). The process of going from the $1 - n$ or nonpermissive state to the n or permissive state is similar to current *activation*, while the opposite direction is similar to current *deactivation*. The rate of change from the nonpermissive state to the permissive state is governed by the variable α_n, while the opposite reaction is governed by the variable β_n. Thus, the rate of change of n with respect to time (dn/dt) follows the differential equation:

$$dn/dt = \alpha_n \cdot (1 - n) - \beta_n \cdot n. \qquad (3)$$

In other words, the change in the proportion of the current that is flowing across the membrane (dn/dt) is governed by the proportion that changes from nonpermissive to permissive ($\alpha_n \cdot (1 - n)$) minus the proportion that changes in the opposite direction ($\beta_n \cdot n$). If we assume that the membrane potential is held steady until there is no change in n (i.e., until $dn/dt = 0$; denoted by n_∞) we find that at steady state

$$\alpha_n \cdot (1 - n_\infty) - \beta_n \cdot n_\infty = 0 \qquad (4)$$

solving for the steady-state activation variable n_∞ we get

$$n_\infty = \alpha_n/(\alpha_n + \beta_n). \qquad (5)$$

The general solution to the first-order differential Equation 3 above is:

$$n = n_\infty \cdot [1 - e^{(-t/\tau_n)}] \qquad (6)$$

where t is time and τ_n is a time constant determined by:

$$\tau_n = 1/(\alpha_n + \beta_n). \qquad (7)$$

Solving for α_n and β_n using Equations 5 and 7 we find that

$$\alpha_n = n_\infty/\tau_n \qquad (8)$$

and

$$\beta_n = (1 - n_\infty)/\tau_n. \qquad (9)$$

Substituting Equations 8 and 9 into Equation 3 yields the relationship:

$$dn/dt = (n_\infty - n)/\tau_n. \tag{10}$$

In typical voltage-clamp paradigms, three variables are defined by exponential relations with membrane potential: the activation variable n_∞, which is the proportion of current that is on at that membrane potential at steady state (in the lack of inactivation), $\tau_{activation}$, which is the time constant of activation of the current and $\tau_{deactivation}$, which is the time constant of deactivation of the current. The Hodgkin and Huxley time constant τ_n is governed by both $\tau_{activation}$ and $\tau_{deactivation}$. In practice, the function τ_n is chosen to best fit the relationship described by both $\tau_{activation}$ and $\tau_{deactivation}$. Since activation and deactivation are opposites, this function tends to appear as a bell-shaped curve, with one end dominated by $\tau_{activation}$ and the other dominated by $\tau_{deactivation}$. Given an accurate description of the steady state activation (n_∞) and kinetics of activation (τ_n), we can use equations, such as Equation 10 above, to simulate on our comptuers the behavior of the current and the axon or neuron.

Currents that Inactivate

Many currents inactivate during constant-voltage steps within their range of activation. This process of inactivation can be described in the same manner as that of activation. Examples of such currents are I_{Na}, I_T, and I_A. The fast Na^+ current underlying the action potential was described by Hodgkin and Huxley with the following equation:

$$I_{Na} = m^3 \cdot h \cdot gNa \cdot (E - E_{Na}), \tag{11}$$

where m is the activation variable and h is the inactivation variable. Like the activation variable m, the inactivation variable h can be in either a permissive or nonpermissive state of current flow (e.g., the channels are either not inactivated, also known as deinactivated, or inactivated). The simple first-order transitions between these two states is represented in the same manner as for n above:

$$1 - h \underset{\beta_h}{\overset{\alpha_h}{\rightleftharpoons}} h, \tag{12}$$

which gives rise to all of the equations shown above. Transition from the permissive state "h" to the nonpermissive state "$1 - h$" is similar to current inactivation, while transition in the opposite direction is similar to removal of inactivation or deinactivation. Remember that *all* gates must be open for ion current to occur. Thus, for the Na^+ channel, the three activation (m) gates and the one inactivation gate must all be open. This dependence on multiple gates is what endows the sodium channels with their complicated opening and closing patterns that result in a total membrane current which activates with a slight delay and then inactivates.

Computer Implementation of the Model

Voltage-Clamp Mode

In voltage-clamp experiments in real neurons, the amount of current injected through the electrode is adjusted so as to keep the voltage of the neuron constant. This is useful for describing the various voltage-dependent and kinetic properties of the underlying current. These data in turn then predict (through models such as this) the subsequent behavior of the neuron in "real life." Here, computer simulation of voltage clamping of a neuron assumed a perfect voltage and space clamp in a neuron in which all of the membrane acts uniformly (i.e., a single-compartmental model) (see Chapter 4 in *Neurobiology*). The model assumes that the neuronal currents are at steady state during the period prior to application of a voltage step. This would be achieved in a real neuron if the membrane potential were held at a particular level for a time that was much longer than the rate of activation, deactivation, inactivation, or removal of inactivation of that current. During the step, the time course of a current that does not inactivate is described by:

$$I_x(t) = gX \cdot (E - E_x) \cdot [n_\infty - (n_\infty - n_{(t-1)}) \cdot e^{(-\Delta t/\tau_n)}], \qquad (13)$$

where X represents some current. The term n_∞ refers to the steady-state value of n, while the term $n_{(t-1)}$ refers to the value of n at the previous time step. The term Δt is the size of the time step used in the model. Calculation of the time course of the current precedes in time steps the size of which are determined by the number of time points in the menu. After each time step, the total ionic current flowing across the membrane is just the arithmetic sum of the individual currents that are on. This membrane current and the voltage is then plotted on the screen after the values are computed for each time step.

Current-Clamp Mode

In current-clamp mode, a constant-current pulse is injected into the cell and the resulting fluctuation in membrane potential is observed. Simulation of current clamp is brought about by making very small increments in time (e.g., 5 μsec), numerically solving for the changes in all of the different currents using a numerical integration technique, calculating the change in membrane potential based upon these changes in current flow, and then proceeding with the new membrane potential to the next time step. In current-clamp mode, the change in membrane potential per unit time ($\Delta V/\Delta t$) is described by:

$$\Delta V/\Delta t = (I_{inject} + I_{NA} + I_K + I_A + I_T + I_L$$
$$+ I_C + I_{AHP} + I_M + I_{Kleak} + I_{NALEAK})/C_n, \qquad (14)$$

where C_n is the total capacitance of the membrane. (In the present model, this capacitance has been set to 0.29 nF.) The values of each of the ionic currents at each time step are calculated as shown above in Equations 1 through 13. The equations used to describe each of the different currents can be found in Huguenard and McCormick (1992), McCormick and Huguenard (1992), and McCormick et al. (1993).

References

Baker, P. F., Hodgkin, A. L., & Shaw, T. I. (1962). Replacement of the axoplasm of giant nerve fibers with artificial solutions. *J. Physiol. (Lond.)*, *164*, 330–354.

Belluzzi, O., & Sacchi, O. (1991). A five-conductance model of the action potential in the rat sympathetic neurone. *Prog. Biophys. Molec. Biol.*, *55*, 1–30.

Brown, D. A. (1988). M-currents. An update. *Trends Neurosci.*, *11*, 294–299.

Connor, J. A., & Stevens, C. F. (1971a). Voltage clamp studies of a transient outward membrane current in gastropod neural somata. *J. Physiol. (Lond.)*, *213*, 21–30.

Connor, J. A., & Stevens, C. F. (1971b). Prediction of repetitive firing behaviour from voltage clamp data on an isolated neurone soma. *J. Physiol. (Lond.)*, *213*, 31–53.

Eccles, J. C. (1957). *The Physiology of Nerve Cells*. Baltimore: Johns Hopkins.

Hille, B. (1984). *Ionic Channels of Excitable Membranes*. Sunderland, MA: Sinaur Associates.

Hirsch, J. C., Fourment, A., & Marc, M. E. (1983). Sleep-related variations of membrane potential in the lateral geniculate body relay neurons of the cat. *Brain Res.*, *259*, 308–312.

Hodgkin, A. L., & Huxley, A. F. (1952). A quantitative description of membrane current and its application to conduction and excitation in nerve. *J. Physiol. (Lond.)*, *117*, 500–544.

Hodgkin, A. L., & Katz, B. (1949). The effect of sodium ions on the electrical activity of the giant axons of the squid. *J. Physiol. (Lond. (Lond.)*, *108*, 37–77.

Huguenard, J. R., Hamill, O. P., & Prince, D. A. (1988). Developmental charges in Na^+ conductances in rat neocortical neurons: Appearance of a slowly inactivating component. *J. Neurophysiol.*, *59*, 778–795.

Huguenard, J. R., & McCormick, D. A. (1992). Simulation of the currents involved in rhythmic oscillations in thalamic relay neurons. *J. Neurophysiol.*, *68*, 1373–1383.

Jahnsen, H., & Llinás, R. (1984a). Electrophysiological properties of guinea-pig thalamic neurons: An in vitro study. *J. Physiol. (Lond.)*, *349*, 205–226.

Jahnsen, H., & Llinás, R. (1984b). Ionic basis for the electroresponsiveness and oscillatory properties of guinea pig neurons in vitro. *J. Physiol. (Lond.)*, *349*, 227–247.

Jasper, H., Rici, G. F., & Doane, B. (1958). Patterns of cortical neuron discharge during conditioned responses in monkeys. In G. E. W. Wolstenhdine & C. M. O'Connor (Eds.), *Neurological Basis of Behavior* (pp. 277–294). Boston: Little, Brown.

Jones, S. W., & Adams, P. R. (1988). The M-current and other potassium currents of vertebrate neurons. In L. K. Kaczmarek & I. B. Levitan (Eds.), *Neuromodulation. The Biochemical Control of Neuronal Excitability*.

Kay, A. R., & Wong, R. K. S. (1987). Calcium current activation kinetics in isolated pyramidal neurons of the CA1 region of the mature guinea-pig hippocampus. *J. Physiol.*, *392*, 603–616.

Kuffler, S. W., Nicholls, J. G., & Martin, A. R. (1984). *From Neuron to Brain* (2nd ed.). Sunderland, MA: Sinauer Associates.

Llinás, R., & Yarom, Y. (1981a). Electrophysiology of mammalian inferior ovilary neurones in vitro. Different types of voltage-dependent ionic conductances. *J. Physiol., 315,* 569–567.

Llinás, R., & Yarom, Y. (1981b). Properties and distribution of ionic conductances generating electroresponsiveness of mammalian inferior olivary neurones in vitro. *J. Physiol., 315,* 569–584.

Mayer, M. L., Westbrook, G. L., & Guthrie, P. B. (1984). Voltage-dependent block by Mg^{++} of NMDA responses in spinal cord neurones. *Nature, 309,* 261–263.

McCormick, D. A. (1992). Neurotransmitter actions in the thalamuc and cerebral cortex and their role in neuromodulation of thalamocortical activity. *Prog. Neurobiol., 39,* 337–388.

McCormick, D. A., & Huguenard, D. A. (1992). A model of the electrophysiological properties of thalamocortical relay neurons. *J. Neurophysiol., 68,* 1384–1400.

McCormick, D. A., Wang, Z., & Huguenard, J. (1993). Neurotransmitter control of neocortical neuronal activity and excitability. *Cerebral Cortex, 3,* 387–398.

McCormick, D. A., & Williamson, A. (1991). Modulation of neuronal firing mode in cat and guinea pig LGNd by Histamine: Possible cellular mechanisms of histaminergic control of arousal. *J. Neurosci., 11,* 3188–3199.

Nicoll, R. A., Malenka, R. C., & Kauer, J. A. (1990). Functional comparison of neurotransmitter subtypes in mammalian central nervous system. *Physiol. Rev., 70,* 513–565.

Nowak, L., Bregestovski, P., Ascher, P., Herbet, A., & Prochiantz, A. (1984). Magnesium gates glutamate-activated channels in mouse central neurons. *Nature, 307,* 462–465.

Shepherd, G. (1994). *Neurobiology* (3rd ed.). New York: Oxford University Press.

Steriade, M., & Deschênes, M. (1984). The thalamus as a neuronal oscillator. *Brain Res. Rev., 8,* 1–63.

Steriade, M., & McCarley, R. W. (1990). *Brainstem Control of Wakefullness and Sleep.* New York: Plenum Press.

Storm, J. F. (1990). Potassium currents in hippocampal pyramidal cells. *Prog. Brain Res., 83,* 161–187.

Yamada, W. M., Koch, C., & Adams, P. R. (1989). Multiple channels and calcium dynamics. In *Methods in Neuronal Modeling. From Synapses to Networks.* Cambridge, MA: MIT Press.

Answers to Study Questions

Determination of Resting Membrane Properties

1. The movements of any particular species of ion (e.g., K^+, Na^+) across the membrane is strongly influenced by the concentrations of that ion on both sides of the membrane and the voltage across the membrane.
2. The equilibrium potential is that potential difference (voltage) across the membrane that just offsets the tendency for the ion to flow down its concentration gradient.
3. Potassium ions are in higher concentration inside the cell than out and therefore tend to flow out of the cell, while Na^+ ions are in higher concentration outside the cell than in and therefore tend to flow into the cell. A negative charge of -100 mV is needed to attract K^+ from exhibiting a net outward flow, while a positive charge of $+41$ mV is needed to repel Na^+ from entering the cell.
4. If the concentrations of K^+ were 3.1 mM inside and 135 mM outside, then the equilibrium potential would be $+100$ mV instead of -100 mV. Likewise, if Na^+ was 31 mM outside and 145 mM inside, then the new equilibrium potential for Na^+ would be -41 mV. Therefore, the new resting potential would be $+65$ mV instead of -65 mV.
5. The membrane potential does not change, since the relative contribution of $pKleak$ and $pNaleak$ to the membrane potential is the same. By doubling $pKleak$ and $pNaleak$, you double the total permeability, and therefore conductance, of the membrane. The steady state of the membrane response to current injection is determined by Ohm's law: V (response) $= I/g$. If conductance (g) is twice as large, then the response (V) will be half as large.

Mechanisms of Action-Potential Generation

1. The upswing of the action potential is mediated largely by Na^+ ions moving down their concentration gradient and into the cell. The downswing is dominated by K^+ ions going down their concentration gradient and therefore exiting the cell.

2. Ionic currents and the underlying ionic channels have two important processes: *activation* and *inactivation*. *Activation* refers to the opening of the activation gate, allowing ions to flow through the channels (if the channel is not inactivated). Deactivation is when this activation gate closes again. *Inactivation* refers to another process whereby the current flowing through the channels is inhibited, even though the activation gates are open. Removing this inactivation is referred to as deinactivation. Channels can only pass ions when they are activated, but not inactivated. In other words, when both gates are open. Inactivation is typically slower than activation. For example, the Na^+ current rapidly activates and more slowly inactivates, thereby generating a brief duration current. Some currents do not inactivate. The K^+ current used in the present model does not inactivate despite maintained depolarization.

3. The inward Na^+ current is kinetically faster than the outward K^+ current (see Figure 9). In addition, the Na^+ current activates at more hyperpolarized levels than does the K^+ current (see Figure 10; the step to -45 mV activates the Na^+ current only). Since the Na^+ current is faster, and since it activates at more hyperpolarized levels, the membrane potential is able to rapidly depolarize in a positive-feedback manner before the exit of K^+ (along with the activation of the Na^+ current) allows for the repolarization of the action potential.

4. At the peak of the action potential, both the Na^+ conductance and the K^+ conductance are active (see Figure 13) and therefore the cell does not reach the equilibrium potential for Na^+.

5. The amplitude of an ionic current is determined by: (1) the maximal conductance in the membrane (i.e., the total number of ionic channels for that current), the amount of the current that is (2) activated but (3) not yet inactivated, and (4) the distance between the membrane potential and the equilibrium potential for the ions underlying that current. As the membrane potential gets close to the equilibrium potential, the amplitude of the current becomes smaller. At the equilibrium potential, there will be no net current flow, no matter how much of the current is activated.

Multiple Ionic Currents in Central Neurons

1. The A-current inactivates, while the "delayed rectifier" K^+ current of Hodgkin and Huxley does not. This inactivation allows the A-current to control the rate of generation of action potentials.

2. A fast K^+ current, known as I_C, that is activated by Ca^{++} entry during

the action potential is important in many cell types for assisting in the repolarization of the action potential. A kinetically slower K$^+$ current, known as I$_{AHP}$, that is also activated by rises in intracellular Ca^{++} levels, is important in controlling the response of the cell to prolonged depolarization. Activation of this current reduces cellular excitability and slows down the rate of action-potential generation.

3. The T-type Ca^{++} current. At hyperpolarized membrane potentials (negative to -65 mV) this current is active and generates a slow Ca^{++} spike that underlies burst discharges. However, at more depolarized membrane potentials, the T-current is inactivated and therefore can no longer generate these slow Ca^{++} spikes.

4. The M-current activates very slowly (e.g., tens to hundreds of milliseconds; see Figure 27), while the "delayed rectifier" K$^+$ current activates much more quickly (e.g., a couple of milliseconds; see Figure 10). This slow rate of activation makes the M-current not particularly useful for action-potential generation. However, the activation of the M-current does selectively decrease the response of the cell to depolarization.

5. Acetylcholine and other modulatory neurotransmitters are known to block the K$^+$ currents I$_{AHP}$, I$_M$, and even some of the resting leak conductance of the membrane, *pKleak*. By reducing these K$^+$ currents, these neurotransmitters can regulate the response and excitability of the cell to other synaptic inputs.

Synaptic Potentials

1. Release of glutamate onto the cell generates excitatory postsynaptic potentials through activation of AMPA and NMDA receptor/ionophores on the postsynaptic membrane. Release of GABA results in the generation of IPSPs through the activation of GABA$_A$ and GABA$_B$ receptors, which are coupled to increases in Cl$^-$ and K$^+$ conductances, respectively.

2. It would be excitatory, since the equilibrium potential for Cl$^-$ would be 0 mV, well above the threshold for action-potential generation (approximately -55 mV).

3. The number of NMDA receptors activated (determined in the modeling program as the conductance of the EPSP: gEPSP), the difference between the membrane potential and the equilibrium potential, and the extracellular concentration of Mg^{++} ions. The NMDA-EPSP becomes smaller as the membrane potential is depolarized towards the reversal potential (0 mV), or as Mg^{++} ions block the NMDA channels, a process that is facilitated by hyperpolarization.

4. They conduct both Na$^+$ and K$^+$ ions and therefore have a reversal potential in-between E$_{Na}$ and E$_K$.

Index—IBM PC Version

Page numbers in *italics* indicate illustrations.

73

---

*Write to Drs. John Huguenard at Department of Neurology, Stanford University Medical Center, Stanford, CA 94305 or David McCormick, Section of Neurobiology, Yale University Medical Center, 333 Cedar St., New Haven, CT 06510.